Models and Measurements of the Cardiac Electric Field

Models and Measurements of the Cardiac Electric Field

Edited by

E. SCHUBERT

Humboldt University of Berlin
Berlin, German Democratic Republic

PLENUM PRESS • NEW YORK AND LONDON

Library of Congress Cataloging in Publication Data

Main entry under title:

Models and measurements of the cardiac electric field.

"Proceedings of the satellite symposium at the Twenty-eighth Interna-
tional Congress of Physiological Sciences on the cardiac field, held July
8–11, 1980, in Dresden, German Democratic Republic"—Verso t.p.
  Bibliography: p.
  Includes index.
  1.Electrocardiography—Congresses. 2. Heart—Models—Congresses. 3.
Electric fields—Measurement—Congresses. I. Schubert, Ernst, Professor
Dr. med. habil. II. International Congress of Physiological Sciences (28th:
1980: Dresden, Germany)
RC683.5.E5M63   616.1'207547   82-5334
                                                          AACR2
ISBN-13: 978-1-4684-4246-5   e-ISBN-13: 978-1-4684-4244-1
DOI: 10.1007/978-1-4684-4244-1

Proceedings of the Satellite Symposium at the Twenty-eighth International
Congress of Physiological Sciences on the Cardiac Electric Field, its
Measuring and Modeling, held July 8–11, 1980, in Dresden, German Dem-
ocratic Republic

© 1982 Plenum Press, New York
A Division of Plenum Publishing Corporation
233 Spring Street, New York, N.Y. 10013

Softcover reprint of the hardcover 1st edition 1982

PREFACE

The electric field of the heart was described diagrammatically
for the first time by A. Waller in 1888. However, it was not until
a little more than ten years ago that with the development of micro-
electronic techniques, it became accessible to biophysical modeling,
to exact physiological measurements, and to application in advanced
clinical diagnosis. These possibilities opened the way to the
treatment of questions which are called the direct and the inverse
solution of the cardioelectric problem.

Several groups of investigators are now working to achieve a
complete biophysical and physiological description of the generation
of the cardiac electric field. This work could well form the basis
for a new method of diagnostic measurements, with applications even
in clinical cardiology, delivering important information by a non-
invasive investigation of the patient.

Several conferences have stimulated international exchange of
the results of research on the cardiac electric field. Among others,
the satellite symposium of the XXV International Congress of Physiological Sciences on the electric field of the heart, in Brussels,
August 2-3, 1971, and the Conference on Measuring and Modeling of
the cardiac electric field, in Smolenice near Bratislava, June 14-
17, 1976, may be considered predecessors of the Dresden symposium
the proceedings of which are presented in this volume.

One interesting result of the continuing research is the increase in clinical investigations using the "mapping" technique of
the electric field of the heart for the improvement of diagnostics
in myocardial infarction, conduction disturbances, etc., as evi-
denced in a remarkable number of contributions. Thus, for the future
we may expect rapid progress in the practical use of the results of
biophysical and physiological investigations on the measurement and
modeling of the cardiac electric field.

E. Schubert

CONTENTS

vii

MEASUREMENTS IN THE CARDIAC ELECTRIC FIELD

CONTENTS

CLINICAL APPLICATIONS OF CARDIAC ELECTRIC
FIELD MAPPING

1. INTRODUCTORY REVIEW

ELECTROCARDIOGRAPHY AND MAGNETOCARDIOGRAPHY TODAY

Pierre Rijlant

Institute Solvay de Physiologie
Free University of Brussels
Brussels, Belgium

The recent developments of magnetocardiography have provided the basis for a valid confrontation of electro-cardiograms and magnetocardiograms. The general tendency is to emphasize the similarity of both recordings, both originating from the same electrical generator system. In his 1979 review, Geselowitz although considering the configuration of the magnetocardiogram's waveform to be much the same as that of the electrocardiogram, also signals the great clinical potential interest in the measurements of currents of injury by magnetocardiography as performed already in the dog's heart and evokes the influence of the movements of the blood on the external magnetic field. This component of the magnetic field, not related to the activation process of the heart, was considered in 1980 at the Paris meeting of the 8th European Congress of Cardiology. A quantitative appraisal of the magnetic signals related to the movements of the heart and of the blood was attempted by Norman Tepley who analyses the component of the magnetic field that is proportional to the ambient field and arises from the presence of diamagnetic blood and tissue. This also makes for the possibility to detect the blood flow in the aorta and in the large arteries. By placing the subject in a large system of Helmholtz coils, Tepley either compensates or increases the effect of the external or earth's field and thus either suppresses or controls the size of the magnetic field of extraneous origin. This can provide for a non invasive mapping of blood flow but provides also the means to isolate the magnetic field due to the flow of current in the heart. Similar results have been obtained by Peters and her co-workers at the physics department of Enschede in the Netherlands and also already in 1971 by Grynspan.

My purpose is not to examine the contributions of the mechanical activity of the heart to the magnetocardiogram, but to consider to

which account the common origin in the electrical generator system of
the heart justifies a similarity of waveform of the electrical and
legimate magnetic recordings.

In their report at the same Paris Congress, Denis and his co-
workers have shown that the magnetic field measured at the surface of
the thorax has a waveform that is quite independent of the nature of
the device used for its recording. They then proceeded to the mapping of the magnetic field simultaneously with a mapping of the electrical field in 98 normal individuals and for 42 pathological cases
and established a clear correspondence between both recordings, the
maxima of the electric and magnetic vectors being obtained for a 90°
spatial axis deviation. They insist that the magnetocardiogram does
not contain more information than the electrocardiogram.

If, for a clinical approximation, electrical and magnetic signs
have a similar diagnostic value, this does not hold for a further
comparison. Already in 1972 Plonsey has clearly defined the essential
differences between both records, the electrocardiogram serving to
determine weighted averages of flux sources, while the magnetocardiogram measures weighted averages of the radial components of vortex
sources, the independence of both records resulting from the relative
independence of flux and vortex sources.

Geselowitz has noted that the magnetic heart vector or dipole is
not the same as the electric heart vector or dipole and also that
different torso boundary effects have to be considered. And this justifies also Geselowitz's opinion that further research is required.
Although no new diagnostic information has been provided in humans,
already experimental results indicate promising possibilities. The
exploration of the magnetocardiogram in ventricular hypertrophy by
Siltanen, Saarinen and their co-workers in Helsinki has shown that
new information of clinical value can be provided by magnetic maps
based on 51 measurements. This already emphasizes the clinical trend
towards magnetic mapping, foreseen at the very start by Cohen in 1967
and 1969. Mapping has also been performed by cardiologists in Rome,
Barbanera and his co-workers, with a relatively simplified equipment
to answer the limitations given to clinical use by technical difficulties.

Although the similarity of both electrical and magnetic recordings, already justified on a theoretical basis by Rush in 1975, provides an apparent justification to some delay in the clinical application of magnetocardiography, a second thought should prevent this
easy escape.

The electrical generator system of the heart is responsible for
both the magnetocardiogram and the electrocardiogram. Nevertheless
important differences have to be considered in the building up of
both fields.

The origin of the technology of modern magnetocardiography calls already for our attention on the location of the actual magnetic sources at a distance of the measuring device. Magnetometers are flown over the ground where possible sources of oil are deeply embedded and record the anomalies of the magnetic distribution. This holds for the magnetocardiogram which explores the whole extent of the heart and not specifically its surface. The components of both electric and magnetic records are necessarily different from one another.

A very simple comparison will also call for the caution that should be exercised. In an isolated nerve, explored in an air chamber, the electrical field has no magnetic counterpart. To provide for a magnetic field the nerve has to be immersed in a conducting fluid such as Locke solution providing for the diffusion in a large volume of fluid of the electrical current flowing at the outside of the fibers. This has been clearly shown by Wikswo, Barach and Freeman in Science. This recalls also that the flow of current from biological sources is in a closed loop, with a very short generator gap, and that the overall flow has a constant level along the whole extent of the path.

In the isolated nerve both currents inside and outside the fibers follow parallel paths with a negligible distance between them and the two magnetic fields cancel one another. But Wikswo's paper also signals some of the advantages provided by the magnetic recording when the cancelling out is reduced by the immersion in a conducting fluid. Not only are valid measurements possible at a distance of the tissue, but they provide a direct appraisal of the current density and compensate for boundary effects. The conclusion of the authors that magnetic measurements are a strong complement to electrical recordings should be seriously considered.

We should also reconsider here the 1968 contribution of David Cohen to magnetic encephalography, who demonstrated outside the human scalp weak alternating magnetic fields produced by α-rhythm currents, as this signals the relative ease with which an overall view on the electrical event can be obtained by magnetic recordings, although the integration of electrical measurements remains more difficult. Recently Malmivuo has also proceeded to a quantitative comparison of the magnetoencephalogram and the electroencephalogram. These papers give in a lead to the more important differences between magnetic and electric signals and emphasize the need for a deeper insight in the componental structure of both recordings.

The electrocardiogram measures potential differences between chosen locations in or on the heart or on the surface of the body. Each measure is only related to the difference in the electrical level of the two locations that are actually compared. Usually either the surface of the heart or of the body are explored and boundary conditions have to be taken into consideration. The electro-

cardiogram corresponds thus to the exploration of an electrical
surface, although we have to remember that the distribution of electrical current parallel to the surface is also dependent on the geometry and on the inhomogeneities of the whole extent of the volume
conductor.

For the magnetic measurements the heart and body are completely
transparent and the magnetometer picks up all the elements of the
magnetic fields produced by whatever currents circulation in the
heart, the relative sizes being controlled inversely by the third
power of the distance. At a few centimeters distance of the surface
the magnetometer explores quite a large surface of the heart and of
the body under similar conditions and in depth quite a large amount
of tissue contributes significantly to the magnetic field.

Thus the influence of the local geometry is reduced and we have
to consider that there is already an important integration of what is
going on in a large area of the explored heart. This could cancel
out small localized events and on the other hand provides at once a
much more global appraisal of the conditions prevailing in a large
section of the heart. But this also implies that currents flowing in
opposite direction can provide for magnetic fields that cancel out
partially and make the appraisal of a surface map more difficult.
This difficulty is of course a function of the distance between the
sites of the flows of current and the location of the measuring
device.

The comparative appraisal of precise electrical and magnetic
maps should be systematically attempted. If this is a relatively
simple problem when only localized surface measurements are considered this does not hold any more when a comparative overall view
of the Paris Congress in 1980 asks for the recording of a magnetic
dipole, eventually in a model, compensating partially or totally for
the body's inhomogeneities. We could of course in a model locate a
dipolar generator providing on the surface of an artificial heart or
an artificial body a distribution of potentials similar to the one
actually observed on the heart's or body's surface, but this does not
take into consideration that the overall flow of current in the heart
at each moment utilizes for each element of the generator system a
closed loop and that the overall magnetic field measured at an
adequate distance should cancel out. The so called equivalent dipolar generator answers only the need for an adequate surface distribution but does not respond to the need for a realistic flow of current in the model and thus the magnetic field measured has no value
whatever for the actual heart's activity.

If some integration of actual magnetic surface measurements or
magnetic maps is considered, this should take into consideration many
dipolar source generators and accept the variation in time of each of
the individual sources.

We should also be very careful in the comparison of magnetic measurements on the isolated heart; either suspended in air or immersed in Locke fluid as this necessarily is to modify very considerably the flow of current in the heart when, as is the case in the living individual, a large conducting volume external to the heart is provided.

REFERENCES

Barbanera, S., Bordoni, F., Carelli, P., Modena, I., and Romani, G.L., 1980, Magnetocardiographic maps using a superconducting magnetometer in an unshielded environment, 8th Europ. Cong. Cardiol. Paris, Abstracts, 1137:90.

Cohen, D., 1968, Magneto encephalography: Evidence of magnetic fields produced by alpha rhythm currents, Am. An. Adv. in Science, 161:784-786.

Denis, B., Machecourt, J., Favier, C., Pellet, J., Wolff, J. E., and Martin Noel, P., 1980, Interêt clinique de la magnetocardiographie; relations entre magnetocardiogramme et electrocardiogramme, 8th Europ. Cong. Cardiol. Paris, Abstracts, 0553:50.

Geselowitz, D. B., 1979, Magnetocardiography. An overview, IEEE Trans. Biomed. Eng., BME, 26:497.

Malmivuo, J. A. V., 1980, Distribution of m.e.g. detector sensitivity - an application of reciprocity, Med. A. Biol. Eng. and Comput., 18:365-370.

Malmivuo, J. A., 1980, Theoretical reasons for selecting VMCG as a standard MCG recording system, 8th Europ. Cong. Cardiol., Paris, Abstracts, 3131:245.

Peters, Maria J., Wevers Henke, J. J., and Van der Marel, L. C., 1980, Magnetocardiography in an external magnetic field, 8th Europ. Cong. Cardiol. Paris, Abstracts, 2317:192.

Plonsey, R., 1963, Reciprocity applied to volume conductors and the ECG, IEEE Transact. BME, 10:9-12.

Plonsey, R., 1972, Capability and limitations of electrocardiography and magnetocardiography, IEEE Trans. BME, 19:239-244.

Rosen, A., and Inouye, G. T., 1975, A study of the vector magnetocardiographic waveform, IEEE Transact. BME, 22:167-171.

Rush, S., 1975, On the independence of magnetic and electric body surface recordings, IEEE Trans. BME, 22:157-167.

Siltanen, P., Saarinen, M., Karp, P., Katila, T., and Varpula, T., 1980, Magnetocardiogram in ventricular hypertrophy, 8th Europ. Cong. Cardiol., Paris, Abstracts, 2369:195.

Tepley, N., 1980, Magneto cardiography and the mapping of blood flow, 8th Europ. Cong. Cardiol., Paris, Abstracts, 0058:10.

Wikswo, J. P., Barach, J. P., and Freeman, J. A., 1980, Magnetic field of a nerve impulse: First measurements, Am. An. Adv. of Science - Science, 208:53-55.

2. MATHEMATICAL AND PHYSICAL MODELLING OF THE CARDIAC ELECTRIC FIELD

HEURISTICAL ALGORITHM FOR THE CORRECT SOLUTION OF THE INVERSE PROBLEM IN TERMS OF THE MODEL OF A CLOSED ELECTRICAL DOUBLE LAYER

Oleg V. Baum

Institute of Biological Physics
Academy of Sciences of the USSR
Pushchino, Moscow Region, USSR

Due to some physico-physiological processes varying with time the distribution of the heart potential over the body surface corresponds to real states of the heart. The inverse problem of the electrocardiological diagnostic task consists in obtaining complete information about the object, i.e. the states of the heart, based on measurements of the distribution of the heart potential in a chosen lead system.

But what does it mean "to know everything about the object"? At first the object has to be described, i.e. a model must be created and its parameters estimated. The optimum criterion proves the adequacy of the direct model to a concrete statement of the inverse problem. The general solution of the inverse problem can be considered as a mapping procedure carried out by means of the algorithm Alg: $\{C\} \rightarrow \{D\}$, where C (c_1, c_2, ... c_r) is a chosen totality of diagnostic signs, D the set of the states of the heart. Most commonly D is the set of diagnoses of an adopted nomenclature. The "empirical way" of solving the inverse problem is based on the algorithms formalizing the medical logic or using specific methods of pattern recognition, which practically do not concern the ECG genesis while the mapping proceeds. A "physical way" of solution is possible, when an "equivalent electrical heart generator" (EEHG) is introduced, which permits obtaining an intermediate characteristic of the electric field of the heart. In this case the mapping procedure is performed in two steps:

$$\{C\} \xrightarrow{Alg_1} \{Parameters\ of\ EEHG\} \xrightarrow{Alg_2} \{D\}$$

The "physical way" of solving the inverse problem offers some advantage, e.g. the relative simplicity of Alg_1 when EEHG structure has

11

been chosen a priori, and the clearness of interpretation of the information obtained. However, there are some limitations, e.g. the lack of a unified opinion on the EEHG structure and the diagnostic value of its components. Besides, it is rather laborious to choose Alg_2 that in essence reduces the problem to the "empirical way". These disadvantages are mainly due to the existing gap between well-developed models of the electric heart activity with the high level of a structural resolution of the object (models of electrogeneous membranes, cellular models, models of formal excitable media) and the integral models at the level of the organ as a whole that are currently applied in cardiodiagnostics (equivalent dipole models, tables of ECG-syndromes) see Titomir (1980), Nelson and Geselowitz (1976), Baum et al. (1973).

In the general case the inverse problem of electrocardiology is inaccurate due to the arbitrariness in the choice of the EEHG structure or possible inaccuracies in the procedure of determining the EEHG parameters of the selected structure according to the ECG measurements as well as to the ambiguity in mapping the set of the calculated generator characteristics in the form of a set of diagnoses. It is of interest, that the viewpoints of physicists and physicians on the inverse and direct problems are opposite: as a rule physicists consider the inverse problem to be inaccurate, but from the viewpoint of the physicians, it is just the electrophysiological direct problem which they assume to be inaccurate. In fact, an inverse problem, in which the state of the heart of a patient is determined by the ECG registration, is easier to be understood by a physician, who by chance may find a simple agreement between the ECG syndrome and the diagnosis. On the other hand, the direct problem is that there is a patient, whose diagnosis is known, e.g. to be within the "norm". Display of his ECG with some lead systems, will be extremely difficult to be solved by the physician. There is no contradiction, since in the physician's mind the diagnostic task from the very beginning is understood not as an inverse problem, but as a direct one. This proves that the choice of the model is crucial because the model adopted must be adequate to the problem stated.

The modelling of the electrical activity of the heart at different structural levels with subsequent unification of the intermediate models into a bio-physically based model of the ECG genesis represents a fruitful approach, that offers considerable scope for an optimum combination of both empirical and physico-physiological means for the solution of the inverse problem (Baum, 1977a). The requirements for such models were reported in 1969 at the Third International Biophysics Congress (Baum, 1969) and in 1972 at the 1st Colloquium on Electrocardiology, where the first results for the realisation of a two-dimensional model were demonstrated too (Baum et al., (1974).

For the models of ECG genesis the conceptual basis is represented by an electrical double layer along the boundary surface S of the

electrically active myocardium. For it the dependency of the intensity of the elementary generators distributed over the heart
$\{U(x,t)\}$ (where $x \epsilon S$) and of the potential $\phi(1,t)$ of the trunk surface
point 1 at the moment t can be written as an integral expression
which, with respect to $\{U\}$, is the Fredholm equation of the first
order:

$$\phi(1)_t = \oint_S K(1,x) \cdot U(x)_t dx. \tag{1}$$

While realizing such a model by a computer, either the requirement to register the ECG-signal in the frequency band limited from
f_{max} (e.g. the choice of the corresponding time quantum Δt) or the
medical requirements for the solution of diagnostic tasks in determining the extensivity of the pathological foci (e.g. the choice of
the linear size quantum Δx) can be a criterion of spatial and temporal discretization. In both cases the quantitative values are related as $\Delta x = v \cdot \Delta t$, where v has the dimension of a velocity.

Experiments with two-dimensional models (Baum, 1977a; Baum et
al., 1974) proved to be advisable for revealing the fundamental po-
tentialities and limitations of the electrocardiographic method of
diagnosis and allowed to gain a certain experience in the realization
of three-dimensional modifications of the models adequate to the
problems of electrocardiology.

Let us consider a matrix equation (2) realizing the model of the
electrical heart activity represented by a piecewise-smooth electrical double layer S situated along the boundaries of the electrogeneous myocardium:

direct problem

$$[k_{1i}] \times [u_{it}] = [\phi 1t] \tag{2}$$

inverse problem

$i = 1,2,\ldots,n;$
$t = 1,2,\ldots,T;$
$1 = 1,2,\ldots,m;$

where u_{it} are the instantaneous potential values of the ith element
of the layer proportional to the value of the transmembrane action
potential (TAP); k_{1i} the transfer coefficients of the ith element to
the lead point 1 determined by the geometry of the heart and the
characteristics of the medium, and ϕ_{1i} the instantaneous ECG value at
point 1 in the system of m leads. All the elements of the surface S
are grouped according to similarity in their TAP-curves designated
$U_j(\tau)$ $(j = 1,2,\ldots,p)$. While modelling the direct problem - in
equation (2) this problem is schematically shown by arrows - the
elements of the matrix $[u_{it}]$ are estimated as follows:
$u_{it} = U_{j(i)}(t-h_i)$, where h_i is the delay of the excitation of a given
element with respect to the beginning of the cardio-cycle, and $j(i)$
is the function of affiliation, which means the distribution of the
curves $U_j(\tau)$ over the layer elements.

The simulation of the direct problem of electrocardiology, e.g. studying the problems of the ECG genesis with the use of a model provides good grounds for the purposeful revealing of diagnostic reserves of the cardiac electrical field in both the "empirical" and the "physical" ways of solving the inverse problem.

In the first case, a model of the electrical heart activity that describes the activity of the heart in terms of current electrocardiology, given in the geometry of the heart and the conducting system, the ratio of velocities of excitation propagation and the shapes of TAP-curves in various sections of the heart reflecting the course of metabolic processes suggests the creation of a model of the pathogenesis of alterations of the potential relief over the trunk. Such a model must elucidate whether there is a probability to get the ECG in a definite shape corresponding to the disease of the patient under investigation. In this case the norms and the pathologies which are of interest to us have to be formulated within the limits of the model parameters. This presumes a close collaboration of physicians, physiologists and biophysicists. All together, these factors provide a radically new approach to the development of the algorithms for the "empirical way", based on the methods of multivariate statistical analysis, e.g. the model can be used as a "teaching sample generator". If the model is adequate and the limits of the normal and of pathological states are given and comprehended in the realm of biophysical and physiological paramaters, we achieve an autoverification of the ECG. This can be generated additionally within the given limits as closely as needed, thus providing representation of the corresponding samples and a possibility for estimating the laws of distribution of the diagnostic signs. Such a generator must lead to more coherent bases and further development of the algorithms simulating the medical logic.

As for the "physical way" of the solution of the inverse problem the EEHG in the form of a piecewise-smooth electrical double layer holds the greatest promise for the systems of automatic cardiodiagnostics of future generations taking into account the progress in adjacent branches of cardiology and a potential level of research efforts. This problem requires the complex creation of physicomathematical and physiological models of the activity of the heart as well as technical analogues directed towards the formulation of biophysically based requirements for such systems (Baum, 1977b; Baum et al., 1978).

To solve the inverse problem in terms of equation (2) means the following: to determine the elements of the matrix $[u_{it}]$ from a given matrix $[k_{li}]$ and measured elements of the matrix $[\phi_{lt}]$. This in turn allows to estimate the delay array $\{h_i\}$ and the arrays $\{U_j\}$ and $\{J_j\}$ reflecting the topography of the propagation of excitation, the shape of the TAP curves and their distribution over S. S means the surfaces of the epicardium and the endocardium as well as the surfaces of probable intramural heterogeneities. In such a case, we obtain

clarity of the interpretation of the results obtained, and the second
step of the diagnostic procedure of choosing the Alg_2 is considerably
simplified. In a certain sense, it is even eliminated. When the op-
portunity arises to use the background experience gained in physiology and biophysics of the heart, we can connect the disturbances in
conduction and metabolic processes in the heart with the model parameters obtained in the course of the solution of the problem. However, the solution of the inverse problems in these terms presents
considerable theoretical and practical difficulties. Since the columns in matrix $[k_{1i}]$ are on definition linearly dependent due to the
closing of the double layer S, so no inverse matrix $[k_{1i}]^{-1}$ exists.
At the same time in specific instances instead of the matrix $[k_{1i}]$
only its approximation $[\tilde{k}_{1i}]$ and the results of the measurements with
an experimental error $[\tilde{\phi}_{1t}]$ are known instead of the accurate values
of ECG recordings.

In order to find appropriate methods for the reduction of the
problem and the provision of its stability at some alterations of the
initial data, it seems advisable to use two- and three-dimensional
models as a "target range" considering, that the wanted solution of
system (2) for the inverse problem is that value of $[u_{it}]$, which
creates a modelling potential relief $[\phi_{1t}]$ in the direct problem.

Using one of the modifications of the model, we studied a heur-
istical method of the inverse problem solution (Baum, 1977c). This
suggested that not the whole surface S becomes excited simultaneously, and the specific characteristics of functions $U_j(\tau)$ are the
following: when $\tau<0, U_j(\tau) = 0$, when $0<\tau<T_j$, $U_j(\tau)$ is a negative
function, the phase "0" of the TAP is approximated by a potential
jump, and $U_j(T_j) = 0$, where T_j is the duration of a corresponding
TAP. For the successive time intervals the equation (2) is solved
with respect to the column matrix $[u_i]_t$ (i = 1,2,...n) on the assumption that $(t - h_i) = \tau_i < 0$ for one or several elements of the surface S. This indicates, that corresponding components of $[u_i]_t$ equal
to 0 are due to the characteristics of $U_j(\tau)$. In other words, we
think that these elements are not yet excited. It should be noted,
that such elements exist on the surface S in the course of the formation of the QRS complex and the P wave for ventricles and auricles,
respectively. This allows us to cancel a corresponding column from
the matrix $[k_{1i}]$ at a given moment. Thus, the columns of the reduced
matrix will not be linearly dependent since the double layer gets
loosened. The only solution of the reduced equation (2) can be described as $u_{it} = \theta_{it} + W(t)$ (i = 1,2,...n), where $[\theta_i]_t$ is the solution sought for, estimated above as a column of the initial matrix
in the direct problem; W(t) is an additive error independent of the
number i. Obviously, there are only two variants of the solution.
If for a given region, e.g. with a number i = q, the hypothesis proposed $(t - h_g<0)$ appears to be incorrect, and in fact $t - h_q= 0$, e.g.
the q-region became excited during the interval Δt between the successive cycles of the direct model functioning, then $W_t = -U_{j(q)}(0)$,

and, hence, each component of $[u_i]_t$, as compared with the moment
(t - 1) will posses a positive spike of about 100 mV. But this con-
tradicts the properties of the function $U_j(\tau)$, as formulated above,
and indicates that the assumption $\tau_q > 0$ is incorrect. Then we ought
to consider the similar hypothesis for another unexcited region.
Here the delay h_q is assumed to be equal to the current value of t.
If the qth region is really not yet excited (e.g. $t - h_q < 0$), then
$W(t) = 0$ and no anomalies occur. Hence at the moment t we have the
right solution $[u_i]_t = [\theta_i]_t$, and the solution of equation (2) should
be continued, assuming that for the next time interval $t + 1 - h_q < 0$,
up to the next anomaly met in the solution, etc. Thus, the topography of the propagation of the excitation $\{h_i\}$ and that part of the
TAP-curves, which corresponds in time to the P wave and the QRS complex are recognized completely. To discover the values of the remaining elements of the matrix $[u_{it}]$ the elements of the surface are
segregated into groups according to the identity of the corresponding
TAP curves. For this purpose, the function j(i) is plotted from the
solution of the equation $[k_{1i}] \times [\hat{A}_i] = [G_1]$ where $\hat{A}_i$ is the area below
the TAP curve in the ith region with the accuracy up to an arbitrary
additive constant independent of i, G_1 is the scalar interpretation
of the intraventricular gradient in the lead 1. Then, according to
the affiliated function j(i) the TAP curves are extrapolated for the
next interval, which is the following QRS period. Thus basing on the
solution obtained in the preceding time interval, the corresponding
elements of the matrix sought for, are filled in. After these procedures, we obtain the unique solution of the equation (2) for the
subsequent time intervals up to the end of the cardiocycle. The results of the model studies show that, when initial data are given
accurately, there are no difficulties in the solution of the inverse
problem in terms of the model proposed.

At the same time, taking into account possible errors in the
initial data of the inverse problem, and in order to obtain a stable
approximate solution $[\tilde{u}_{it}]$ with an accuracy corresponding to that of
the target of $[\tilde{k}_{1i}]$ (Robb et al., 1979) and $[\tilde{\phi}_{1t}]$, the equation (2)
is solved for consequent values of t by the method of regularization.
This reduces the problem to a search for the column $[u_i]_t$ which provides some functions $F([u_i]_t, \alpha)$ with a minimal value, where α is the
regularization parameter (Tikhonov, 1965). To work out the methods
of the models under the conditions of really achievable accuracy of
the initial data, the algorithm for the inverse problem solution was
realized as a digital computer program. This program is joint to
that one realizing the model of the ECG genesis with altering the
input data, e.g. for simulation of various "patients" and different
measurement accuracy with the aim to choose the optimum values of α
for the range of alterations simulated.

The next step is to work out the algorithms for the solution of
the inverse problem in respect to the real objects and to develop the
methods allowing to gain correct information of the improvement of

the matrix $[\tilde{k}_{1i}]$. This step requires the creation of a physiological model, as a test bench providing vital activity in the course of the perfusion of the hearts of warm-blooded animals and allowing to take measurements required for the solution of direct and inverse problems.

REFERENCES

Baum, O. V., 1969, Model of the heart electrical activity, III International Biophysics Congress, Cambridge, p. 157.

Baum, O. V., Dubrovin, E. D., and Titomir, L. I., 1973, Biophysical approach to electrocardiography and the problem of "clearness" of interpretation of ECG-information, in: "Modelling and automated analysis of ECG", Nauka, Moscow, p.35, (in Russian).

Baum, O. V., Kiselev, E. E., and Orlova, L. I., 1974, Use of the physic-mathematical models for some electrocardiological problems, in: Neue Ergebnisse der Elektrokardiologie, E. Schubert, ed., Fischer, Jena, p. 47.

Baum, O. V., 1977a, Modelling of the heart electrical activity, in: Biophysics of complex systems and radiation disturbances, Nauka Moscow, p. 119.

Baum, O. V., 1977b, Problems and perspectives of utilization of the models for solving the direct and inverse problems of electrocardiology, in: Proceedings of the 30th ACEMB, Los Angeles, p. 4.

Baum, O. V., 1977c, Problems of models utilization for the inverse problem solving, in: "Theory and practice of automatization of the electrocardiological and clinical investigations, Kaunas, p. 239 (in Russian).

Baum, O. V., Orlova, L. I., and Popov, L. A., 1978, Problems of automatization of electrocardiological investigation. System approach, in: "Modern Electrocardiology", Z. Antalóczy, ed., Akad. Kiado, Budapest; Excerpta Medica, Amsterdam, p. 243.

Robb, R. A., and Ritman, E. L., 1979, High speed synchronous volume computed tomography of the heart, Radiology, 133:655.

The theoretical basis of electrocardiology, 1976, C. V. Nelson and D. B. Geselowitz, eds., Clarendon Press, Oxford.

Tikhonov, A. N., 1965, On incorrect problems of the linear algebra and the stable method of their solving, Dokl. Akad. Nauk SSSR, 163:591 (in Russian).

Titomir, L. I., 1980, Electrical heart generator, Nauka, Moscow (in Russian).

THE INVERSE POTENTIAL PROBLEM APPLIED TO THE

HUMAN CASE *

P. Colli Franzone, L. Guerri, B. Taccardi **
and C. Viganotti

Institute of Numerical Analysis of C.N.R.
Pavia, Italy
** Institute of General Physiology
University of Parma, Italy

INTRODUCTION

In this paper, we discuss the critical points that are encountered in extending a numerical procedure for the computation of epicardial from surface potential values to the human case.

In our approach to the inverse potential problem (Colli Franzone et al., 1979) the mathematical problem, here briefly recalled, was attacked via regularization techniques and numerically approximated using the finite element method. In vitro collected data from a normal beat of an isolated dog heart were used to validate the numerical methods (Colli et al., 1978).

A generalized singular value decomposition algorithm was implemented to reduce the computational load of the inverse procedure and to make it feasible to apply the methods to large sets of data.

With the aim of achieving the best accessible estimation of epicardial potentials, the following points relating to the numerical inverse procedure were debated in account to effectiveness of different regularization methods, influence of the distance of the surface from the epicardium; accuracy of the transfer matrix, and criteria for choosing the smoothing parameter on the basis of surface data.

---

*   This work was performed in the frame of the Special Program on Biomedical Engineering of the Italian National Research Council (C.N.R.).

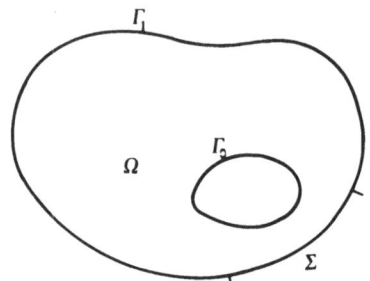

Fig. 1.   A general picture of the problem geometry

Emphasis was placed on this last point which represented the greatest obstacle to be overcome before solving the inverse problem in man. We propose here an empirical criterion which, even if not yet fully explained at the theoretical level, proved to be very effective on the in vitro data.

Results are reported relating to epicardial maps evaluated at a few significant moments of a normal beat in a human subject.

MATHEMATICAL FORMULATION OF THE INVERSE POTENTIAL PROBLEM

Let us denote by:

Γo   a surface enclosing the heart
Γ_1 the body surface
Σ the torso surface
Ω the body volume between Γ_0 and Γ_1
V(x) the cardiac electric potential at location x$\epsilon\,\Omega$, at time t
k(x) the electrical conductivity of the medium
z(x) the measured potential distribution on Σ, at time t.

Assuming the torso potential values z(x) are known, then the inverse potential problem consists in estimating the "epicardial" potential values on Γ_0 by solving the following Cauchy problem for an elliptic operator of second order:

<u>find</u> V(x) on Γ_0 where V satisfies:

div k(x) grad V(x) = 0 in Ω; $\dfrac{\partial V(x)}{\partial n}$ = 0 on Γ_1;

$$V(x) = z(x) \text{ on } \Sigma$$

It is well known that this problem is ill-posed, i.e. small perturbation on the data on Σ yield "greatly" amplified perturbations in the solution. There is no way to avoid this difficulty since the measured data z(x) are inevitably affected by a noise level and the numerical procedures introduce an approximation and round-off errors.

METHODS OF NUMERICAL APPROXIMATION

To overcome the ill-posedness, the inverse problem is approximated by means of stable problems using regularization techniques (Tykhonv, A.N. and Arsenline, V., 1977). We are then led to solve the following minimization problem dependent on a smoothing parameter $\varepsilon > 0$:

find $u_\varepsilon \epsilon \, \mathcal{U}$: $J_\varepsilon(u_\varepsilon) = \inf\limits_{v \epsilon \mathcal{U}} J_\varepsilon(v)$ where

$$J_\varepsilon(v) = \int_\Sigma |Av-z|^2 d\sigma + \varepsilon \int_{\Gamma_o} |Bv|^2 d\sigma \qquad (3.1)$$

The operator A denotes the observation operator and is defined by:

$$Av = y(v)\Big|_\Sigma$$

and $y(v)$ satisfies the following mixed boundary value problem:

div $k(x)$grad $y(v) = 0$ in Ω; $\dfrac{\partial y(v)}{\partial n} = 0$ on Γ_1;

$$y(v) = v \text{ on } \Gamma_o \qquad (3.2)$$

The regularization technique depends on the choice of the function space $\mathcal{U}$ and of the regularization operator B. Two different cases have been theoretically investigated and applied (Colli et al., 1979), namely:

$\mathcal{U} = H^1(\Gamma_o)$ and $B = \nabla$ (gradient operator)

$\mathcal{U} = H^2(\Gamma_o)$ and $B = \Delta$ (Laplace-Beltrami operator)

By using the method of finite elements, problem (3.1) is approximated by a finite dimensional problem. Its numerical solution is carried out in three steps, hereafter briefly outlined.

Step 1

Construction of the transfer matrix T between Γ_o and Σ which approximates the operator A, i.e. T is the matrix relating epicardial to torso potentials.

Let u be the vector of nodal values of epicardial potentials and z the vector of nodal values of surface potentials. The vectors u and z are bound by the relation:

z = Tu

If m and n are the dimensions of z and u, respectively, due to the linearity of (3.2), the construction of T requires the solution

of n problems related to the discrete three dimensional variational
form of the mixed problem (3.2). T is an ill-conditioned matrix reflecting the ill-posedness of the inverse problem.

<u>Step 2</u>

Solution of the minimization problem:

$$J_\varepsilon(u_\varepsilon) = \min_{v \in R^n} J_\varepsilon(v)$$

where $J_\varepsilon(v) = \dfrac{1}{m} \| Tv-z \|^2 + \varepsilon \| Bv \|^2$

B is the matrix approximating either one of the regularization
operators ∇ or Δ.

The least square problem related to J must be solved for a
sequence of vectors z corresponding to different instants of the heart
beat and, for each moment, for several values of ε. A fast and effi-
ent solution of these problems was achieved by means of the Generalized Singular Value Decomposition applied to the matrices T and B
(Golub, G., and Reinsch, C., 1970; Van Loan, C., 1976). The GSVD
algorithm is a stable algorithm for building two orthogonal matrices
U and V of order m and one invertible matrix X of order n, such that

$$U*TX = D_T$$

$$V*BX = D_B$$

U* and V* denote the transposes of U and V and D_T, D_B are non-negative
diagonal matrices with the diagonal elements arranged in decreasing
order. It is then possible to express the solution u_ε with explicit
dependence upon ε, as follows:

$$u_\varepsilon = Xq_\varepsilon \quad \text{and} \quad q_{\varepsilon,i} = \frac{t_i \, c_i}{t_i^2 + m\varepsilon b_i^2}, \quad i = 1, \ldots, n$$

with $D_T = \text{diag}(t_1, t_2, \ldots, t_n)$

$D_B = \text{diag}(b_1, b_2, \ldots, b_n)$

and $c = U*z$

<u>Step 3</u>

Application of an empirical estimator for the choice of the
smoothing parameter :

<u>find</u> the smallest positive maximum of the function:

$$C(\varepsilon) = \left\| Bu_\varepsilon \right\|^2 + 2\varepsilon \, \frac{d}{d\varepsilon} \left\| Bu_\varepsilon \right\|^2$$

where u_ε solves the least square problem of step 2.

The difficult problem of choosing that smoothing parameter which realizes the best tradeoff between fitting of surface information and smoothing of the solution is related to the nature and to the a priori known information of the overall noise affecting the surface data. If a good approximate value σ of the standard deviation of the noise is known, then a satisfactory choice of ε would be that one which is realized (Gordonova, V.I., and Morozov V.A., 1973) through the equation:

$$\frac{1}{m} \left\| Tu_\varepsilon - z \right\| \simeq \sigma$$

If σ is unknown but the assumption of white noise holds, statistical estimations of ε, recently reported in (Wahba, G., 1977; Golub, G., et al., 1978; Anderssen, B., and Bloomfield, P., 1974; Colli et al., 1979), proved to be very effective.

However, in the actual experimental inverse problem, the overall noise affecting surface data, even if quite low, is not normally distributed and non-stationary because of the interpolation errors introduced by the time-alignement of non simultaneous measurements. Hence, we looked for an estimation of ε that would work without knowledge of σ and with no assumptions about the noise nature and proposed the criterion reported above which we shall call Composite Residual and Smoothing Operator (CRESO).

VALIDATION OF THE NUMERICAL METHODS

In the validation of the methods, the data collected in experiments on isolated dog hearts were used (Taccardi, B., et al., 1972; Taccardi et al. 1976). From 156 potential values collected on the surface of a tank, a cylinder with a diameter of 23 cm and 15 cm high, "epicardial" maps were inversely computed at 108 sites lying on ideal surfaces enclosing the heart at an average distance from the epicardium of 0.5 and 1 cm, respectively, and compared to the correspondingly measured "epicardial" maps.

Comparison was extended to a set of 62 time instants selected from a normal beat of the isolated dog heart.

The accuracy of reconstructed maps was evaluated computing the mean square root relative error and the correlation coefficient between the measured and the computed inner maps.

In the numerical calculations, the domain Ω between heart and body surface was discretized by a three-dimensional mesh of hexahedral elements obtained, subdividing Ω by 12 horizontal sections, each of which was divided into 12 sectors and each sector radially into 8 (4, 3) parts.

An extensive investigation was carried out to evaluate the influence of different factors on the accuracy of inversely computed epicardial maps: stabilization method used, distance of the surface from the "epicardium" and accuracy of the transfer matrix. Under all conditions that value was assigned as the smoothing parameter which we may call optimal, that yielded best fitting of measured and computed epicardial potentials. Moreover, on the same set of experimental data and under all conditions the performance of the empirical estimation of the smoothing parameter was evaluated and compared to the accuracy yielded by the optimal smoothing parameter. Out of all the numerical investigations, here we report only those concluding remarks, which had influence on the design of the numerical experiment in its application to the human.

Stabilization Method

When the inner surface lied 0.5 cm from the epicardium, the ill-conditioning of the transfer matrix T, measured by the condition number of T, i.e. by the ratio between the greatest and the smallest singular value of the matrix, was about $2,2.10^{10}$. Due to this order of ill-conditioning, the numerical results obtained by using the optimal smoothing parameter, showed a weak sensitivity of the best accessible accuracy to the choice of the regularization operator. In fact, the gradient operator yielded slightly better accuracy than all other stabilization methods (see Colli et al., 1979) and was used in the successive numerical tests.

Distance of the Epicardium from the Surface

The measured "epicardial" maps at 0.5 and 1 cm from the epicardium were highly correlated. As an average on the entire normal beat, the correlation coefficient (CC) was over 94%.

In inverse calculations we observed that the CC between measured and computed inner maps at 1 cm were more accurate than those computed 0.5 cm closer to the epicardium. As an average on the entire normal beat, CC was 77% at 1 cm and 65% at 0.5 cm. Since at 1 cm the numerical results were more accurate and the experimental data were highly correlated to those collected at 0.5 cm, most of the successive tests were referred to the data collected at 1 cm from the epicardium.

Mesh Refinement

In reducing progressively the number of the threedimensional mesh elements from 1200 to about 500 by decreasing the number of radial parts per sector, it was observed that the accuracy was practically insensitive to a reduction of the mesh refinement. Therefore the mesh including about 500 elements was choosen in the following, allowing a corresponding reduction of computation time and memory occupation.

This lack of sensitivity to the mesh refinement was attributed to the fact that the numerical approximation errors were of a smaller order of magnitude in respect to the overall noise level affecting the surface data. This conclusion was supported by the fact that decreasing the mesh fineness no increase of surface residuals was observed.

Smoothing Parameter

Under all different conditions which we considered above, the performance of the CRESO criterion we proposed for a choice of the smoothing parameter was also evaluated and the accuracy of reconstructed maps using CRESO was compared to the optimal accuracy.

In all conditions, CRESO proved to be very effective as it is illustrated in Fig. 2.

The CRESO criterion failed in a time interval of about 30 msec between the end of QRS and initial ST, when the ratio between the magnitude of epicardial and surface signals was about 4.5, i.e. higher than the average value on the entire beat (≈ 2.5). This higher ratio is mainly due to the overlapping of the effects produced by the simultaneous presence of depolarization and repolarization events on the ventricles. One example of inverse maps obtained using CRESO at 1 and 0.5 cm from the epicardium is reported on Figs. 3(a) and 3(b), respectively.

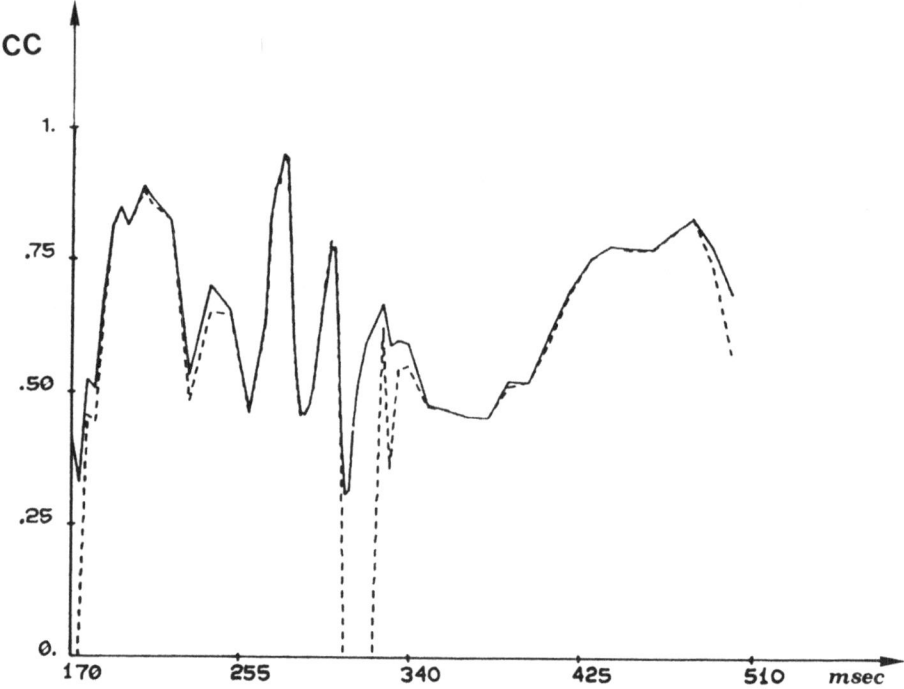

Fig. 2.   Solid line represents the correlation coefficient between
          the measured and the computed epicardial maps when the
          optimal smoothing parameter is used;   dashed line gives
          the same CC when the CRESO criterion is used to estimate
          the smoothing parameter.   In the time interval when CRESO
          failed to work, CC was set to zero.   This result relates
          to the inner geometry at 0.5 cm from the epicardium.

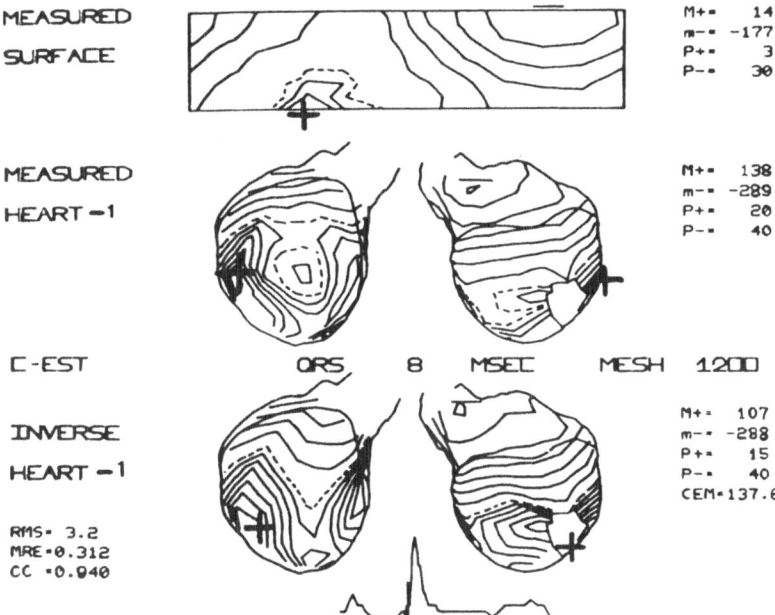

Fig. 3(a). At 8 msec from QRS onset, surface potentials (top) show
          an anterior maximum and a posterior minimum. At 1 cm
          from the epicardium, both the measured (centre) and
          computed (bottom) maps show two separate maxima with
          negativity wedging itself in between. The correlation
          coefficient between measured and computed potentials is
          94%.

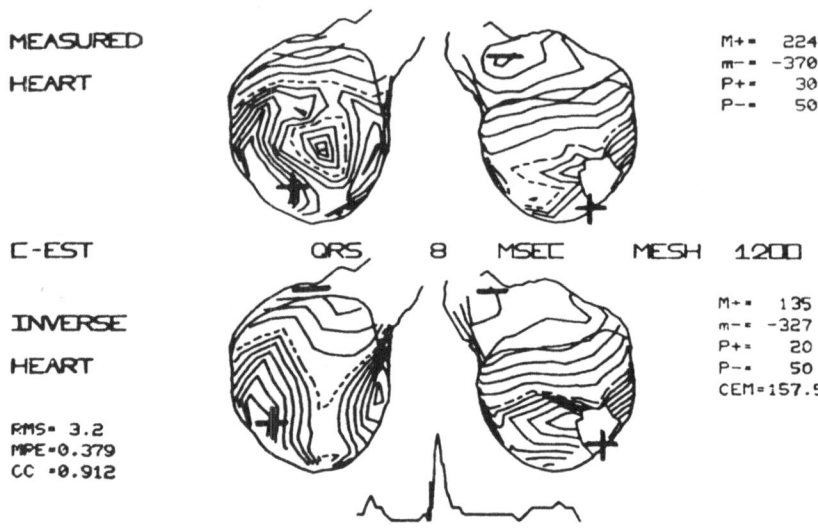

Fig. 3(b). Same time instant as in Fig. 3(a) except that inner potent-
          ials were collected 0.5 cm closer to the epicardium. The
          correlation coefficient between measured and computed
          potentials is now 91%.

We remark that the solution of the problem of choosing the
smoothing parameter on the basis of surface data with no assumptions
on the nature of the noise affecting the data, even if only at an
empirical level, constituted a significant advancement toward the
application of the inverse procedure to the human case.

APPLICATION TO A NORMAL HUMAN SUBJECT

We present here the results obtained on a normal human subject
by extrapolating the procedures we developed and validated on the in
vitro collected data. Extrapolation relates to the fact that, in
first instance, inhomogeneities were not taken into account.

Input information were the geometries of the torso and of the
heart. The first was obtained by a plaster cast of the torso of the
normal subject on which the 230 sites of measurements were marked;
the heart geometry was deduced combining frontal and lateral X-ray-
images of the chest of the same subject and sections of a normal

heart reported on Eycleshymer and Shumaker tables. This information
was used to build an automatic three-dimensional mesh of the body
volume between heart and body surface (Fig. 4(a) and 4(b)). The
meshing, including about 1500 hexahedral elements and 1900 nodes, was
then used to build the transfer matrix T between the 230 sites of
measurement on the body and about 160 locations on the epicardium.
Sites of measurement were among the nodes in the mesh with no need of
making interpolations to get the potential values at the nodes.

An a priori analysis of the ill-conditioning of the transfer
matrix in the human torso showed that the singular values of T decayed
at a slower rate than in the case of the dog heart experiment. The
ratio between the greatest and smallest singular values of T was about
10^6.

The surface potentials were collected from a normal human subject
by a 240-probe instrument (Cottini et al., 1972).

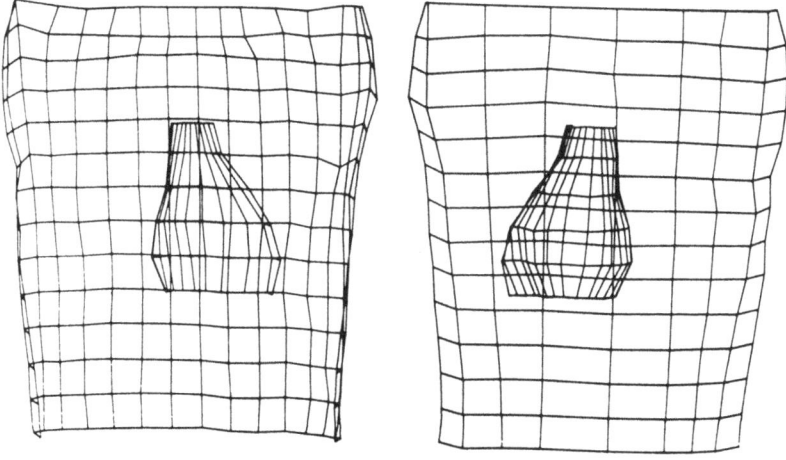

Fig. 4(a). Anterior and posterior torso and the mesh on the body
          surface. The profile of the heart imbedded in the
          thorax is also shown.

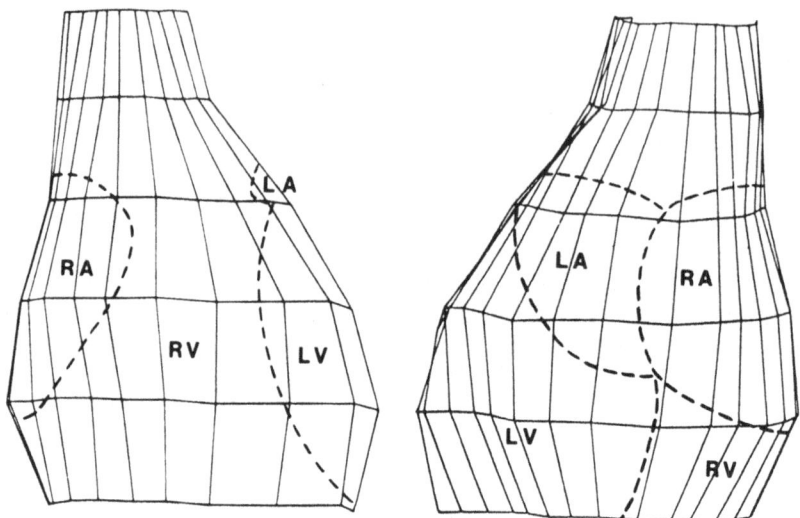

Fig. 4(b). Closer view of the heart: dashed lines indicate the
          approximate locations of atria and ventricles as they
          were viewed anteriorly and posteriorly in the human
          subject.

     By the inverse procedure we discussed above and using the CRESO
criterion to estimate the smoothing parameter, epicardial maps were
computed at 40 moments selected from the entire heart beat. For lack
of space, we are only able to include into this paper one example of
surface map and of the correspondingly computed epicardial map (Fig.
5(a) and 5(b)). In general, the reconstructed epicardial maps showed
patterns in good agreement with the present knowledge of cardiac
electrophysiology. In many cases, as in the example reported here,
epicardial maps indicated wavefront-locations which were hardly pred-
ictable from the features of the corresponding body surface maps.
Note that the surface maps of the normal human subject we considered
showed almost always a dipolar or quasidipolar pattern. In spite of
this, the important events, like the spreading of two separate wavefronts during atrial excitation, the signs of right and left ventricular breakthrough, the excitation of the crista supraventricularis
and of pulmonary infundibulum in the QRS, as well as the events of
atrial and ventricular repolarization were quite evident on the epicardial maps. To what extent body volume inhomogeneities would have
affected these patterns if they were taken into account in the calculations, remains to be investigated.

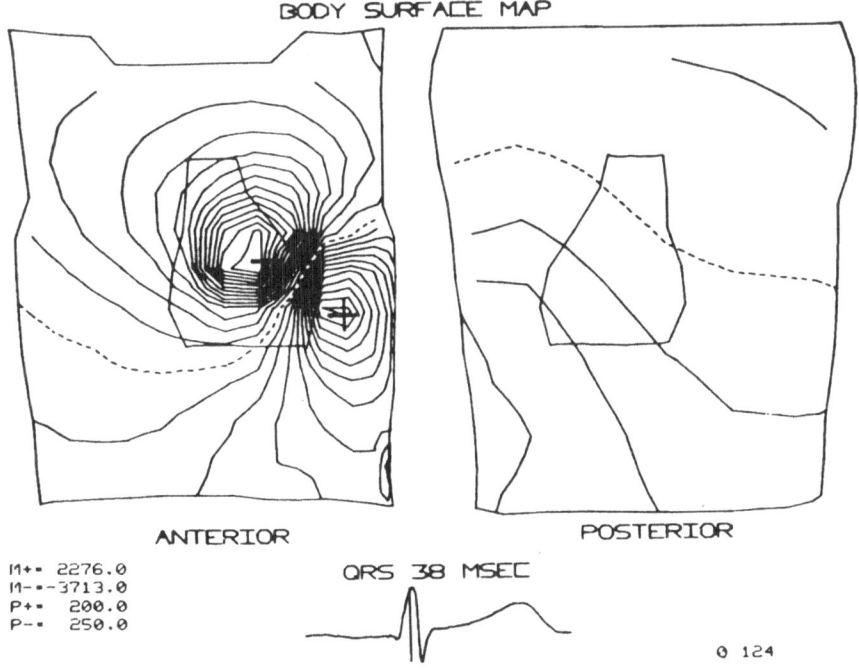

BODY SURFACE MAP

ANTERIOR   POSTERIOR

M+= 2276.0
M-=-3713.0
P+= 200.0
P-= 250.0

QRS 38 MSEC

0 124

Fig. 5(a). At 38 msec from QRS onset, body surface potentials show
a minimum which reached the midsternal area and a maximum
which started its travel to the left.

        Now that epicardial maps in the human may be computed
from body surface maps, the problem of their investigation on the
theoretical and clinical level must be faced. It is hoped that the
more detailed information content of epicardial maps and the close-
ness of this information to the cardiac sources will make the interpretation and classification problems easier to attack than from the
corresponding body surface maps.

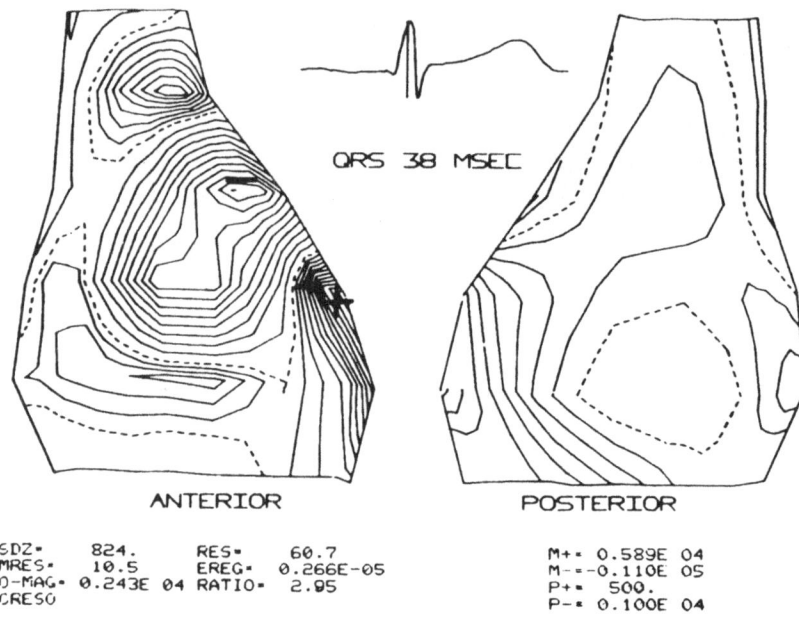

REFERENCES

Anderssen, B., and Bloomfield, P., 1974, Numerical differentiation
    procedures for non-exact data, <u>Numer. Math.</u> 22:157
Colli Franzone, P., Guerri, L., Taccardi, B., and Viganotti, C., 1978,
    A regularization method for inverse electrocardiology applied to
    data from an isolated dog heart experiment, <u>in</u>: "Modern Electro-
    cardiology", Ed. Z. Antaloczy, Amsterdam, Excerpta Medica, pp. 75
Colli Franzone, P., Guerri, L., Taccardi, B., and Viganotti, C., 1979,
    The direct and inverse potential problems in electrocardiology.
    Numerical aspects of some regularization methods and application
    to data collected in isolated dog heart experiments, <u>L.A.N. of
    C.N.R.</u>: 222
Cottini, C., Dotti, D., Gatti, E., and Taccardi, B., 1972, A 240-probe
    instrument for mapping cardiac potentials, <u>in</u>: "The Electrical
    Field of the Heart", Ed. P. Rijlant, Bruxelles, Presses Acadé-
    mique Européennes, pp. 99

Golub, G., and Reinsch, C., 1970, Singular value decomposition and
    least squares solutions, Num. Math. 14:5

Golub, G., Heath, M., and Wahba, G., 1977, Generalized cross-validat-
    ion as a method for choosing a good ridge parameter, Report STAN-
    CS-77-622, Comp. Sc. Dept., Stanford University

Gordonova, V.I., and Morozov, V.A., 1973, Numerical parameter select-
    ion algorithms in the regularization method, Zh. Vycisl. Mat. i
    Mat. Fiz. 13:539

Taccardi, B., Musso, E., and De Ambroggi, L., 1972, Current and poten-
    tial distribution around an isolated dog heart, in: "The Electri-
    cal Field of the Heart", Ed. P. Rijlant, Bruxelles, Presses
    Académique Européennes, pp. 566

Taccardi, B., Viganotti, C., Macchi, E., and De Ambroggi, L., 1976,
    Relationships between the current field surrounding an isolated
    dog heart and the potential distribution on the surface of the
    body, in: "Adv. Cardiol.", 16, Karger, Basel, pp. 72

Tykhonov, A.N., and Arsenine, V., 1977, Solution of ill-posed problems
    J. Wiley & Sons, New York

Van Loan, C., 1976, Generalizing the singular value decomposition,
    SIAM J. Numer. Anal. 13-1:76

Wahba, G., 1977, Practical approximate solutions to linear operator
    equations when the data are noisy,
    SIAM J. Numer. Anal. 14:651

COMPUTER SIMULATION OF CARDIAC EXCITATION

I. Ruttkay-Nedecký, V. Szathmáry, P. Chlebus,
A. Ruttkay-Nedecká

Institute of Normal and Pathological Physiology,
Centre of Physiological Sciences, Slovak Academy
of Sciences, Bratislava, CSSR

Three years ago, the fiftieth anniversary of Einthoven's death was commemorated in Leiden by a symposium on developments in Electro-cardiography during the past 50 years. The topic of computer model studies of the activation of the heart, discussed by Arntzenius et al. (1978) was introduced into the program and this might be regarded as a sign of, let us say, an official acceptation of this line of research as a new development in Electrocardiology.

A promising family of such models are the so called multiple dipole models developed along the lines set by Selvester and coll. (Solomon and Selvester, 1971; Ritsema van Eck, 1972; Arntzenius et al., 1978). They are highly detailed multisegment models with con-straints imposed on the spatio-temporal distribution of generators according to anatomical data and electrophysiological measurements of the spread of activation in isolated, as well as in "in situ" hearts. These models begin to show their utility not only for the exploration of the detection limits of ECG criteria for myocardial hypertrophy (Holt et al., 1969), myocardial infarction (Sylvester et al., 1967), or conduction disorders, but also as an aid to electro-cardiographic diagnosis and last but not least as an alternative to laboratory animal experimentation, and as a tool for performing ex-periments which cannot be realized in vivo by present day techniques.

An example of the last mentioned application is the simulation of the sequence of the myocardial activation in the model of the iso-lated interventricular septum. Till now, this model has been utili-zed for studies of the effect of interindividual variability of left bundle branch geometry on septal activation (Ruttkay-Nedecký, 1979) and for evaluation of the effect of simulated myocardial infarction in the septum on the vectorial parameters of the excitation spread (Ruttkay-Nedecký et al., 1980).

The computer model of the septum is shaped like a half of a thick walled bisected hollow right circular cone. It consists of 5558 evenly distributed points serving as a medium for modelling the myocardial activation sequence. Each point, which represents a current dipole, is either at rest, or activated. It is supposed to be "off" when the excitation front is outside the corresponding region of the heart and "on" after the front has passed through the region. The points follow the all or none law, no spatial or temporal summation takes place. Activation spreads from one point to another and each point is supposed to activate each of its neighbouring points after a short delay, the duration of which reflects the conduction velocity. The myocardial excitation is considered to be the same in all directions. It is to be noted that the direction of the fibers seems to affect the spread of excitation only slightly. Its influence is probably cancelled out because of the variation in fiber directions in subsequent myocardial layers.

The propagation of the excitation takes place in discrete time-steps. The actual duration of these steps is chosen so as to match the computed duration of the activation process with the duration of interventricular septum activation.

Our digital computer simulation program is a somewhat simplified modification of the Huygens' geometrical method of wavefront construction in a three-dimensional isotropic and homogeneous region. The logic employed in generating the propagating front across the set of evenly distributed points in a Cartesian coordinate system is based in our studies on finding the boundary "on"-points following several schemes:

1. The simplest variant is to allow the excitation to propagate from the initial point up to the distance of 5 units, where 1 unit is the distance between two neighbouring points. The boundary surface is then a rough approximation of a spherical surface. The process is repeated then taking the boundary points as starting points.

2. A better variant is to let the excitation front go first to the distance of the square root of three units, then to the distance of the 1 unit, and to repeat this second step once again. This procedure reaches a better approximation of the spherical surface.

3. The best variant was found by looking after combinations of not less than three and not more than five steps, where as distances could be used either one, or the square root of two, or the square root of three. A special program was designed to find out the combinations leading to the least quadratic error with respect to the spherical surface. Twelve combinations of five steps in each case were found to be the most effective ones. One of them was chosen for actual computation: one, one, square root of two, square root of three and one.

After finding the boundary-points, a resultant vector is computed by vectorial summation of unit vectors originating from the
actual boundary points and oriented towards the neighbouring "off"-
points, that means not yet excited points. The process is then
started over again using the last "on"-points - boundary points - as
the centers for the next step.

The output data are presented as graphics, showing the activation front on cuts of the model along Cartesian coordinates, as well
as in the form of the resulting activation vector projections onto
the principal planes of the coordinate system. The simulation program runs either on a Siemens 4004/150, or on an IBM 370/148 computer. Fortran IV was used.

The variability of left bundle branch ramification plays an important role in the temporospatial evolution of the activation fronts
in the myocardium. In earlier simulation studies of the activation
sequence (Lynn et al., 1967; Selvester et al., 1971; Rittsema van
Eck, 1972) the activation was initialized over areas of the septum
derived from electrophysiological data of Durrer et al. (1966) obtained from a restricted number of revived and perfused human hearts.
In all these model studies the same source served for defining an
idealized geometry of input points and imposed heavy restrictions regarding to the study of the role of interindividual variability.

We attempted to approach this problem by taking anatomical
rather than electrophysiological data as basis for defining the geometry of input points in our simulation experiments.

Since the fibrous sheet of the conducting system offers a high
resistance to the electric current, we may suppose that myocardial
excitation starts from areas where specific histologic techniques
fail to visualize the continuity of the bundle branches. It was
assumed therefore that the areas of transition from the morphologic-
ally distinguishable network of the left bundle branch to the ordinary myocardium are the starting points of the electrical activation
fronts invading the septum. Data published by Demoulin and Kulbertus
(1976) were utilized for definition of 49 normal variants of left
bundle branch ramifications and starting points of septal activation.

In an attempt to quantify the component sources of the interindividual variations of instantaneous cardiac vectors recorded from
the body surface, the extent of the variability of the 10 ms - septal
- vector, which might be caused by interindividual differences of the
left bundle branch geometry, was studied with the aid of the described computer model. In each simulation experiment S_i, at each
instant of time T_{1-8}, resultant vectors were computed reflecting the
actual spatial configuration of the activation front. The coordinates of the vector at time T_1, corresponding to the 10 ms vector
were noted. The model of the septum was oriented by tilting the apex

of the half cone from the vertical position 45 degrees to the front
and 45 degrees to the left, the inner surface facing the left side of
the imaginary subject. Thereafter the whole group of the individual
resultant vectors, computed at T_1, was found to be oriented within the
limits obtained in a group of healthy subjects of both sexes, aged 18
- 50 years (n = 79), with the McFee-Parungao axial lead system, as
shown on Fig. 1 and Fig. 2. A comparison of the spread of measured
and of simulated values led to the estimate that about 20 per cent of
the total normal interindividual variability of the 10 ms QRS septal
vector orientation in the left sagittal projection and 13 per cent in
the horizontal projection might be ascribed to interindividual variability of the geometry of the left bundle branch ramification
(Ruttkay-Nedecky, 1979).

In the next step, the same model was used to answer the question,
whether there is any connection between certain anatomical types of
ramification and the spatial orientation of the 10 ms vector. Figure
3 shows three types of ramifications identified from the study of
Demoulin and Kulbertus (1976). The first type is characterized by
three subdivisions, the anterior subdivision is slim and short. The
second type lacks the central subdivision, the posterior one is dom-
inant. The third type is again without distinct central radiation,
there is an intermingling of anterior and posterior radiations.
While the x, y, z components of the computer simulation resultant
vectors, equivalent to the 10 ms QRS vector, had a normal distribution, those belonging to the type 1 ramification with the slim and
short anterior subdivision, were found to be significantly more often
oriented to the left half of the spread of values. Taking also into

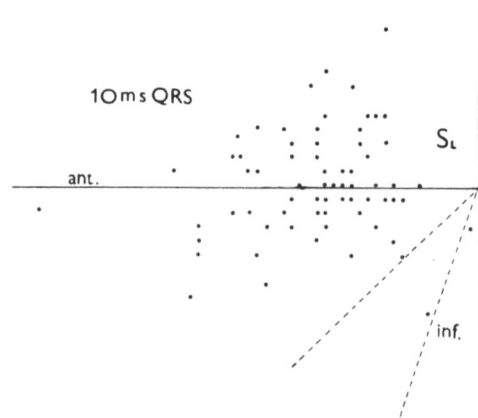

Fig. 1.   Normal distribution of 10 ms QRS vector end-points in the
          left sagittal plane. The sector delineated by broken lines
          indicates the spread of computer simulation values.

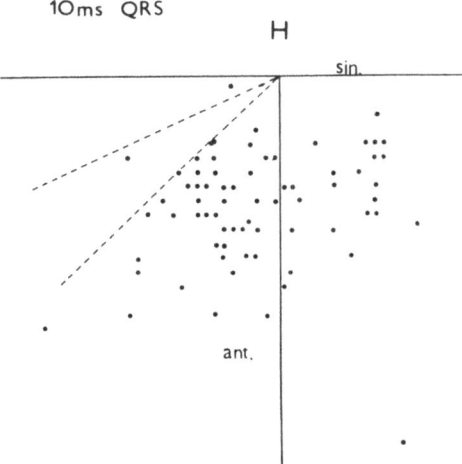

Fig. 2. Normal distribution of 10 ms QRS vector end-points in the horizontal plane. The sector delineated by broken lines indicates the spread of computer simulation values.

Fig. 3. Three types of left bundle branch ramifications, identified on hand of data reported by Demoulin and Kulbertus (1976).

account the physiological rotation of the septum around its long axis in addition to the sagittal and horizontal plane corrections of the long axis orientation of the septum model as described above the orientation of vectors belonging to the slim and short anterior sub-division variety is expected to be reflected in the 12 lead ECG as missing Q waves in leads I, aVL, V_5, V_6. This variety of the ECG therefore might be the expression of a specific anatomical variety of the left bundle branch ramification and not necessarily needs to be a sign of septal fibrosis as described by Burch and de Pasquale (1960).

Another area of electrocardiologic research, where computer sim-ulation of myocardial activation may be of value, is the still con-troversial topic of diagnosing small intramural lesions of the myo-cardium.

Twenty points, abstracted from the above mentioned data and re-presenting a mean normal pattern of starting points for the spread of excitation were selected on the left ventricular surface of the septum model. Four inexcitable areas were simulated, equivalent to about 9, 81, 1125 and 3087 cubic millimeters of tissue. The two smallest areas were localized intramurally. It could be shown, that already the smallest inactive area - 1 point in the model, equivalent to about 9 mm^3 of tissue - affected the temporospatial evolution of the wave-front. The magnitude of the resultant vector was found to be diminished with respect to its normal time course during a small fraction of about 6 ms of total septal excitation time, this change being located at the beginning of activation, at the time of the Q wave of the surface ECG. The ECG surrogate should be a notch, or a slurring.

REFERENCES

Arntzenius, A. C., Schipperheyn, J. J., Huisman, H. P., Kulbertus, H. E., Ritsema van Eck, H. J., Simoons, M. L., Vinke, R. V. H., 1978, Model Studies on Activation of the Heart, Europ. J. Cardiol. 8: 261-270.

Burch, G. E., De Pasquale, N., 1960, A Study at Autopsy of the Relation of Absence of the Q Wave in Leads I, aVL, V$_5$ and V$_6$ to Septal Fibrosis, Am. Heart J., 60:336-343.

Demoulin, J. C., Kulbertus, H. E., 1976, Pathological Findings in Patients with Left Anterior Hemiblock, in: Vectorcardiography 3, North-Holland publ. Co., Amsterdam - Oxford, 220 pp.

Durrer, D., van Dam, R. T., Freud, G. E., Janse, M. J., Meijler, F. L. Arzbaecher, R. C., 1970, Total Excitation of the Isolated Human Heart. Circulation, 41:899-912.

Holt, J. H., Barnard, A. C. L., Lynn, M. S., 1969, A Study of the Human Heart as a Multiple Dipole Electrical Source. I. Normal Adult Male Subjects. II. Diagnosis and Quantitation of Left Ventricular Hypertrophy, Circulation, 40:687-714.

Lynn, M. S., Barnard, A. C. L., Holt, J. H., Sheffield, L. T., 1967, A Proposed Method for the Inverse Problem in Electrocardiography, Biophys. J., 7:925-945.

Ritsema van Eck, H. J., 1972, Digital Computer Simulation of Cardiac Excitation and Repolarization in Man. "Thesis", Dalhousy University, Halifax, Nova Scotia, quoted after Arntzenius et al., 1978.

Ruttkay-Nedecký, I., 1979, Model Study of Septal Activation Starting from the Left Bundle Branch, in: "Progress in Electrocardiology", P. W. Macfarlane, ed., Pitman Medical, Tunbridge Wells, 115-117.

Ruttkay-Nedecký, I., Szathmáry, V., Chlebus, P., Národa, J., Ruttkay-Nedecká, A., 1980, Modelling of the Myocardial Infarction in the Interventricular Septum, in: IV. Krajowa Konferencja Biocybernetika i inzynieria biomedyczna, Tom I. Polska Akademia Nauk, Poznań: 185.

Selvester, R. H., Kalaba, R., Collier, C. R., Belleman, R., Kagiwada, R. H., 1967, A Digital Computer Model of the Vectorcardiogram with Distance and Boundary Effects: Simulated Myocardial Infarction. <u>Amer. Heart J.</u>, 74:792-814.

Selvester, R. H., Palmersheim, J., Pearson, R. B., 1971, VCG inverse Model for the Prediction of Myocardial Disease, <u>in</u>: Vectorcardiography 2, I. Hoffman, ed., North-Holland publ. Co., Amsterdam - London: 706 pp.

Solomon, J. C., Selvester, R. H., 1971, Myocardial Activation Sequence Simulation, <u>in</u>: Vectorcardiography 2, I. Hoffman, ed., North-Holland publ. Co., Amsterdam - London: 706 pp.

ABOUT THE ACCURACY OF THE METHOD OF CALCULATING INTEGRAL

CHARACTERISTICS OF THE CARDIAC ELECTRICAL GENERATOR

M. Maco and P. Kneppo

Institute of Measurement and Measuring Technique
of the Slovak Academy of Sciences
Bratislava, Czechoslovakia

At the present time owing to an inaccuracy and a considerable subjectivity in the evaluation of contemporarily used medical methods many investigators try to propose more accurate methods containing more excessive informations about the state of the heart. Therefore we would like to suggest a method for the construction of suitable consistent physical models of the electrical activity of the human heart.

At the moment several authors make the proposal to model the equivalent electrical generator of the heart as a system of current dipoles. In this model each dipole represents the activity in a particular region of the heart. Parallel efforts have been devoted to another equivalent generator, to the multipole generator. This method has first been proposed by Geselowitz (1960, 1965) and many authors (e.g. Titomir, 1975) have afterwards continued in this direction. A group of investigators from the Institute of Measurement and Measuring Technique of the Slovak Academy of Sciences in Bratislava has also derived a system of formulas relating surface potentials measured over the human thorax with dipolar and quadrupolar components of the equivalent multipole generator and suggested a method for the calculation of these parameters. The set of those parameters was named the integral characteristics (Kneppo and Titomir, 1979). The dipolar components of the equivalent generator are expressed in the formulas:

$$\begin{aligned} p_x &= \gamma \int \phi \cdot dS_x \\ p_y &= \gamma \int \phi \cdot dS_y \\ p_z &= \gamma \int \phi \cdot dS_z \end{aligned} \qquad (1)$$

and the quadrupolar components of the equivalent generator are expressed in the form:

$$A_{20} = \gamma \cdot \int \phi \, (2z \cdot dS_z - x \cdot dS_x - y \cdot dS_y) \quad B_{20} = 0$$

$$A_{21} = \gamma \cdot \int \phi \, (x \cdot dS_z + z \cdot dS_x) \quad B_{21} = \gamma \cdot \int \phi \, (z \cdot dS_y + y \cdot dS_z) \tag{2}$$

$$A_{22} = \frac{\gamma}{2} \cdot \int \phi \, (x \cdot dS_x + y \cdot dS_y) \quad B_{22} = \frac{\gamma}{2} \cdot \int \phi \, (y \cdot dS_x + x \cdot dS_y)$$

where ϕ is the potential measured at chosen points over the human thoracic surface, x, y, z are the carthesian coordinates of the measuring points and γ is the mean conductivity of the human thorax. Potential expressions for the calculation of the coordinates of the electrical center of the heart were received by minimizing the quadrupolar in the form of:

$$X_{ect} = \frac{1}{4p} \cdot \left[-A_{20} c_x (1 + c_z^2) + 2A_{21} c_z (2 - c_x^2) - 2B_{21} c_x c_y c_z + 2A_{22} c_x (4 - c_x^2 + c_y^2) + 4B_{22} c_y (2 - c_x^2) \right]$$

$$Y_{ect} = \frac{1}{4p} \cdot \left[-A_{20} c_y (1 + c_z^2) - 2A_{21} c_x c_y c_z + 2B_{21} c_z (2 - c_y^2) - 2A_{22} c_y (4 + c_x^2 - c_y^2) + 4B_{22} c_x (2 - c_y^2) \right] \tag{3}$$

$$Z_{ect} = \frac{1}{4p} \cdot \left[+A_{20} c_z (3 - c_z^2) + 2A_{21} c_x (2 - c_z^2) + 2B_{21} c_y (2 - c_z^2) - 2A_{22} c_z (c_x^2 - c_y^2) - 4B_{22} c_x c_y c_z \right]$$

where $\quad p = \sqrt{p_x^2 + p_y^2 + p_z^2}$

is the magnitude of the dipolar moment vector and

$$C_x = \frac{p_x}{p}$$

$$C_y = \frac{p_y}{p}$$

$$C_z = \frac{p_z}{p}$$

are the direction cosines of this vector.

Formulas (1) and (2) are expressed in an integral form. But data obtained from measurements are to be evaluated by a digital computer. Therefore they have to be transformed into a discrete form, i.e. the integration has to be converted into a summation. An infinitesimal small element of the human thoracic surface dS is replaced by the real triangular surface ΔS, where the tops of the triangles are neighbouring points of measurement of the potentials. Potential ϕ then is the arithmetical mean of the potentials from the tops of the triangles and x, y, z are the coordinates of the triangle centroid entering the formulas for calculation of the quadrupole components (2) as the coordinates of new calculated "points of measurement".

The accuracy of such an approximation is inversely proportional to the area of the triangle. That means the accuracy of this method is proportional to the density of the measuring points. Because of practical reasons the number of points of measurement cannot exceed a certain limit. It follows from above that some error is caused by the triangular approximation of the human thorax. Value and course of that error have been determined by a computer test.

Okada (1956) derived formulas for the calculation of the surface potential distribution over a homogeneous conducting cylinder from a current dipole placed at an arbitrary point inside the cylinder and oriented arbitrarily:

$$V_z = \frac{-2p_z}{ab\pi\gamma} \sum_{n=1}^{\infty} \sin\frac{n\pi z'}{a} \cdot \cos\frac{n\pi z}{a} \sum_{k=1}^{\infty} \frac{I_k\left(\frac{n\pi\rho'}{a}\right)}{I_{k-1}\left(\frac{n\pi b}{a}\right) \ I_{k+1}\left(\frac{n\pi b}{a}\right)} \cdot \cos k(\Phi-\Phi')$$

$$V_\rho = \frac{2p_\rho}{ab\pi\gamma} \sum_{n=1}^{\infty} \cos\frac{n\pi z'}{a} \cdot \cos\frac{n\pi z}{a} \sum_{k=1}^{\infty} \frac{\frac{ka}{n\pi\rho'} \cdot I_k\left(\frac{n\pi\rho'}{a}\right) + I_{k+1}\left(\frac{n\pi\rho'}{a}\right)}{I_{k-1}\left(\frac{n\pi b}{a}\right) + I_{k+1}\left(\frac{n\pi b}{a}\right)} \cdot \cos k(\Phi-\Phi') \qquad (4)$$

$$V_\varphi = \frac{2p_\varphi}{ab\pi\gamma} \sum_{n=1}^{\infty} \cos\frac{n\pi z'}{a} \cdot \cos\frac{n\pi z}{a} \sum_{k=1}^{\infty} \frac{\frac{ka}{n\pi\rho'} \cdot I_k\left(\frac{n\pi\rho'}{a}\right)}{I_{k-1}\left(\frac{n\pi b}{a}\right) + I_{k+1}\left(\frac{n\pi b}{a}\right)} \cdot \sin k(\Phi-\Phi')$$

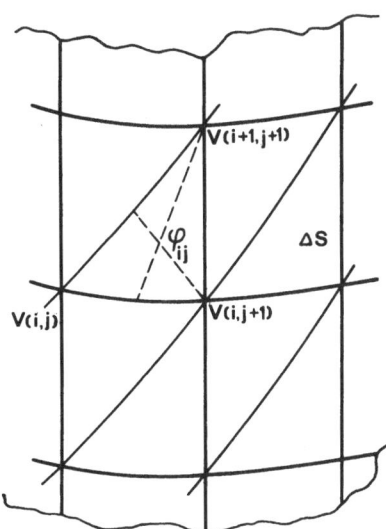

Fig. 1. Method of triangulation of human thoracic surface

where a- means the height of the cylinder, b- is the radius of the
cylinder, γ is the electrical conductivity of the cylinder. The
primed cylindrical coordinates ρ', ϕ', z' indicate the dipole position and ρ, ϕ, z are the coordinates of the measuring points. p_z, p_e
and p_ϕ are dipole components in the cylindrical coordinate system.
I_k is the k'th order modified Bessel function of the first kind. The
cylinder is surrounded by a non-conducting dielectric medium. Since
the medium of the cylinder is linear, the superposition of the three
solutions yields a solution for an arbitrarily oriented dipole.

During our computer test the current dipole of a known constant
moment is moved along a circular trajectory arbitrarily located inside the cylinder (Fig. 2). At chosen points of the trajectory the
dipole gradually changed its orientation by small steps resulting in
360° as a whole. After this gradual change of orientation the dipole
also changes its position and moves to the next point of the trajectory. By means of Okadas' (1956) formulas we calculated the surface
potential distribution over the whole cylinder for each position and
for each orientation of the dipole. Then by means of the above mentioned triangulation of the cylinder surface and by application of
expressions for the calculation of integral characteristics (1), (2),
(3) we calculated the dipolar and the quadrupolar components of the
equivalent electrical generator and the coordinates of the trajectory
of the electrical center.

By comparing the original input data (dipole components and coordinates of points of trajectory) with the calculated data an estimation of the error of the triangular approximation of the human
thorax surface became possible.

Following from preliminary results of our computer test it is
allowed to claim, that the error of the method of calculation of the
integral characteristics possesses only a multiplicative character
and is not essentially influencing either the shape of the trajectory
of the electrical center of the heart or the shape of the trajectory
of the end point of the dipole moment. Therefore both the multipole
model of the electrical activity of the human heart and the method of
calculation of integral characteristics are suitable for further application in the analysis of the electrical field of the human heart.

It seems to be necessary to perform more detailed analyses of
the influences of both the number of points of potential measurements
and the choice of positions of those points on the accuracy of the
method for calculation of the integral characteristics of the equivalent electrical generator of the human heart.

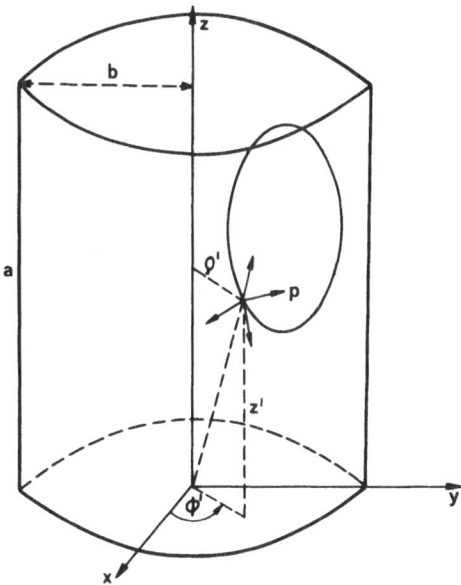

Fig. 2.   Circular cylinder showing cylinder constants, coordinate
system and dipole trajectory and orientation.

REFERENCES

Geselowitz, D. B., 1960, Multipole Representation for an Equivalent
        Cardiac Generator, Proc. IRE, 48:75-79.
Geselowitz, D. B., 1965, Two Theorems Concerning the Quadrupole Ap-
        plicable to Electrocardiology, IEEE Trans. Biomed. Eng. BME,
        12:164-168.
Kneppo, P., and Titomir, L. I., 1979, Integral Characteristics of the
        Human Cardiac Electrical Generator from Electric Field Meas-
        urements by Means of an Automatic Cylindrical Coordinator,
        IEEE Trans. Biomed. Eng., BME, 26:21-28.
Okada, R. H., 1956, The Image Surface of a Circular Cylinder, Am.
        Heart J., 51:489-500.
Titomir, L. I., 1975, Integralnyje charakteristiki električeskoj
        volny vozbuždenija serdca, Biofizika, 20:693-698 (in Russian).

AN ATTEMPT TO LOCALIZE THE PRE-EXCITATION SITE IN WOLFF-PARKINSON-WHITE PATIENTS BY MEANS OF A MATHEMATICAL MODEL

E. Macchi*, L. Guerri**, B. Taccardi***,
V. Bonatti****, A. Rolli****, G. Botti****

INTRODUCTION

Normal conduction of excitation between the atria and the ventricles takes place through the atrio-ventricular (A-V) node, the His bundle and its branches in approximately 80 msec. In the Wolff-Parkinson-White (W.P.W.) syndrome the excitatory process reaches the ventricular myocardium earlier than normal (pre-excitation) through an anomalous conducting bundle which bypasses the A-V node somewhere along the fibrous rings (Kent bundles) or emerges from the His bundle or the initial portions of the bundle branches (Mahaim fibers). Accordingly, pre-excitation spreads to the ventricles at different sites in individual patients, thus bringing about different potential patterns at the body surface. Since excitation spreads also through the normal pathways, a mixed beat occurs.

Initially the anomalous excitation wavefront occupies a small region of ventricular myocardium and in this early phase the instantaneous electrical source may be approximated by a dipolar generator. Locating the pre-excitation dipole by means of non-invasive methods

---
*
  Istituto di Matematica, Università di Parma e
  Istituto per le Applicazioni del Calcolo, C.N.R., Roma
**
  Istituto di Analisi Numerica, C.N.R., Pavia
***
  Istituto di Fisiologia Generale e Centro SIMES,
  Università di Parma
****
  Divisione di Cardiologia, Ospedale Maggiore, Parma

is clinically important in order to limit epicardial mapping to a
small area during surgical correction of this anomaly.

In a previous investigation (De Ambroggi et al., 1976) body
surface maps relating to ventricular excitation enabled the probable
location of the pre-excited area to be identified in 42 patients.
Further reports from other groups confirmed the usefulness of body
surface maps in classifying WPW patients (Yamada et al., 1974;
Spach et al., 1978).

In this study an attempt was made to determine the position and
the moment of the equivalent dipole associated with the early spreading of pre-excitation, by means of a mathematical model. Many mathematical models currently used in inverse electrocardiology do not
suit this problem, since they represent the electrical activity of
the heart by means of a number of dipoles fixed in position and direction but varying in intensity (Barnard et al., 1976). More recently, a single moving dipole representing cardiac electrical activity
for normal and ectopic beats has been localized in a homogeneous
torso model of a dog (Savard et al., 1980). Other models provide an
indirect solution by computing the potential distribution on a surface
closely surrounding the heart (Barr et al., 1976; Colli et al., 1979).
In the present study a three-dimensional human torso model has been
developed, which takes into account inhomogeneities due to lungs and
intracavitary blood. The mathematical problem is of the inverse type:
it consists in identifying the position and moment of a dipole generating body surface maps which are in good agreement with the measured
ones during pre-excitation. The numerical solution is obtained as a
sequence of direct problems in which the position and moment of a
dipole are changed according to a suitable strategy until the difference between computed and measured surface potentials is minimized.
A number of simplifications enabled the model to be implemented on a
small computer without requiring excessive computation time.

METHODS

Body Surface and Intracardiac Measurements

A 45-year old patient who reported episodes of paroxysmal supraventricular tachycardia and one episode of paroxysmal atrial fibrillation was considered as case study. Surface ECG recordings revealed
a shortened PR interval (110 msec), a prolonged QRS interval (140
msec) and type A pre-excitation patterns with positive delta waves in
leads I, II, aVL and VI to V6. Intracavitary ECG recordings localized
the pre-excited area in the posterior paraseptal ventricular wall
close to the left A-V ring. Body surface maps were recorded by apply-
ing 219 electrodes to the anterior and posterior chest surface by
means of vertical rubber straps. Electrocardiograms were simultaneously obtained from all the electrodes by using an automated 240-
channel instrument with on-line amplification, multiplexing and

analog-to-digital conversion at 120 kHz. The data relating to one
second of cardiac activity were trapped in a dynamic shift memory and
stored on magnetic disk or tape. Real time averaging of 128 successive beats was also performed in order to improve the signal-to-noise
ratio. The electrode bearing rubber straps were then replaced by
adhesive paper straps with a mark at each electrode position. A
transparent mold of the torso was rapidly made by using the 3M Light-
cast II Casing System. After the electrode positions were marked,
the mold was removed and the coordinates of the 219 lead points were
measured. Chest X-ray and data from an anatomy textbook (Eycleshymer
and Schoemaker, 1911) were used to reconstruct the shape of lungs and
intracavitary blood within the thorax. The outside surface of the
torso as well as the internal surface were then discretized by a
finite number of surface elements (Fig. 1).

Mathematical Model

The mathematical procedure consisted in determining within the
torso model the position and the moment of that dipole which minimized
the difference between computed and recorded body surface potentials.
To achieve this purpose the following functional relationship must be
minimized

$$\int_{S_T} \left[V_R(Q) - V_C(Q,P,\underline{m}) \right]^2 . dS$$

where S_T is the external torso surface, V_R is the recorded potential
distribution at points Q of S_T, V_C is the computed potential distribution at points Q of S_T due to an internal dipole located at point
P and with moment $\underline{m}$. The integral is to be computed on S_T. The
optimal dipole is determined by successive approximations.

The potential V_C is computed as the solution of the forward problem of electrocardiology stated as follows. In a homogeneous and
isotropic medium (torso) other homogeneous and isotropic subregions
(lungs and intracavitary blood) with different conductivities are im-
bedded. Let S_L and S_H be the surfaces which bound the lungs and the
blood cavities respectively: S_T then contains S_L and S_H. The potential field of a dipole at point P and with moment $\underline{m}$ inside S_T satisfies Poisson's equation

$\Delta V_C(Q) = \delta(Q-P).\underline{m}$, where δ is the Dirac delta function.

The normal component of the electric field must be continuous
across S_L and S_H but zero on S_T, since the torso surface is insulated.
The problem can be transformed into an equivalent one described by an
integral equation (Barnard et al. I, II, 1967). In either form the
solution of the problem can be obtained only by numerical methods because of the complex geometry and the inhomogeneities of conductivity.

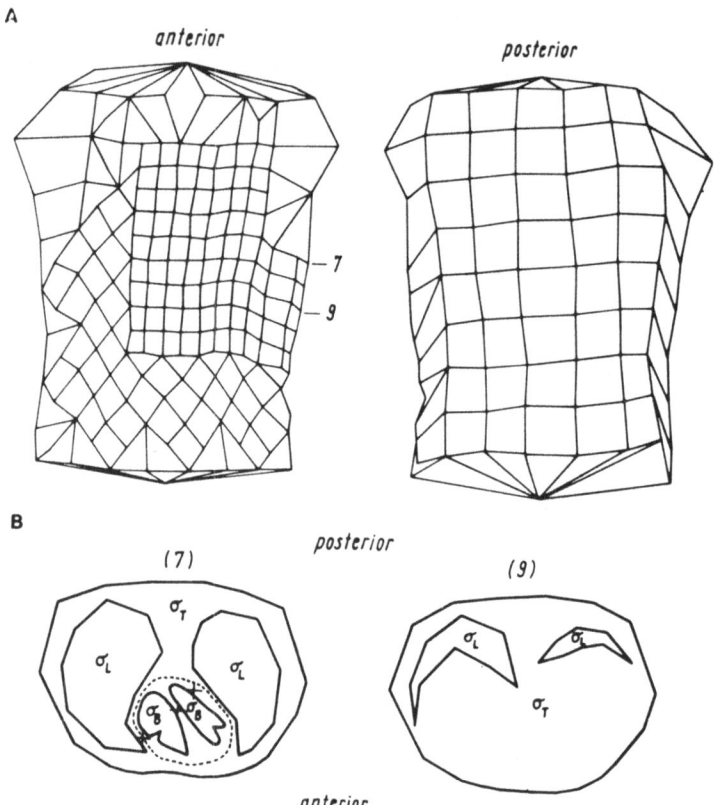

Fig. 1.   Outside surface of the torso of the WPW patient:
A: anterior and posterior projection of the surface
elements. B: horizontal projection of levels 7 and 9 of
the thorax containing lungs and intracavitary blood;
σ_T, σ_L and σ_B represent the different conductivities of
the thorax, lungs and blood respectively ($\sigma_T=1$, $\sigma_L=0.24$,
$\sigma_B=3.2$ in arbitrary units).

The optimal dipole is computed as the solution of a sequence of forward problems related to the different positions and moments of the test dipole. The search region may be greatly reduced by using the constraint, that the dipole location may be found only along the A-V fibrous rings. The search region is then discretized with a set of points distributed along the rings (Fig. 2). Let F denote this region. For a given dipole at point P and with moment $\underline{m}$ the discretized form of the integral equation approach gives rise to a system of linear equations whose solution yields, among other results, the potential values on the surface elements of S_T. Let n_F be the number of points of the set F which discretizes the A-V rings and n_T, n_L, n_H the number of surface elements of S_T, S_L and S_H respectively. For each point P_K, K=1, ..., n_F in F we compute the vectors $\underline{a}_K$, $\underline{b}_K$, $\underline{c}_K$ of dimension n_T, which represent the potential on the surface elements of S_T generated by three unitary dipoles located in P_K and parallel to the reference axes. In this way we obtain the transfer coefficient matrices A, B, C of dimension n_T . n_F. These matrices are computed only once and depend on the geometry of the surfaces S_T, S_L, S_H and the conductivity coefficients. A dipole located in P_K with moment $\underline{m} = (\alpha, \beta, \gamma)$ generates on S_T a potential $\underline{V}_C = \alpha\underline{a}_K + \beta\underline{b}_K + \gamma\underline{c}_K$. If $\underline{V}_R$ denotes the recorded potential on S_T the residual between computed and recorded potential on S_T is $\underline{r} = \underline{V}_C - \underline{V}_R$. For a fixed K the mean relative error (MRE) is

$$|\underline{r}|/|\underline{V}_R| = |\underline{V}_C-\underline{V}_R| / |\underline{V}_R| = \left[\sum_1^{n_T} (V_{Ci}-V_{Ri})^2 / \sum_1^{n_T} V_{Ri}^2 \right]^{1/2}.$$

This function depends on $\underline{m} = (\alpha, \beta, \gamma)$. The moment $\underline{m}^+ = (\alpha^+, \beta^+, \gamma^+)$, which minimizes the MRE, defines the dipole at P_K, which gives the best fitting to $\underline{V}_R$. Let ϕ_K denote the minimum MRE at P_K. Then the optimal dipole is located at that point of F, which yields the smallest value for ϕ_K.

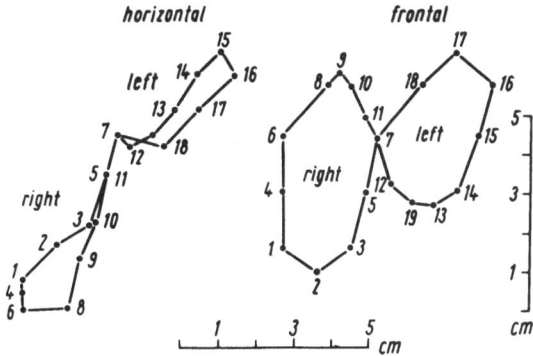

Fig. 2.   Horizontal and frontal projection of the A-V fibrous rings discretized with a set of 19 points.

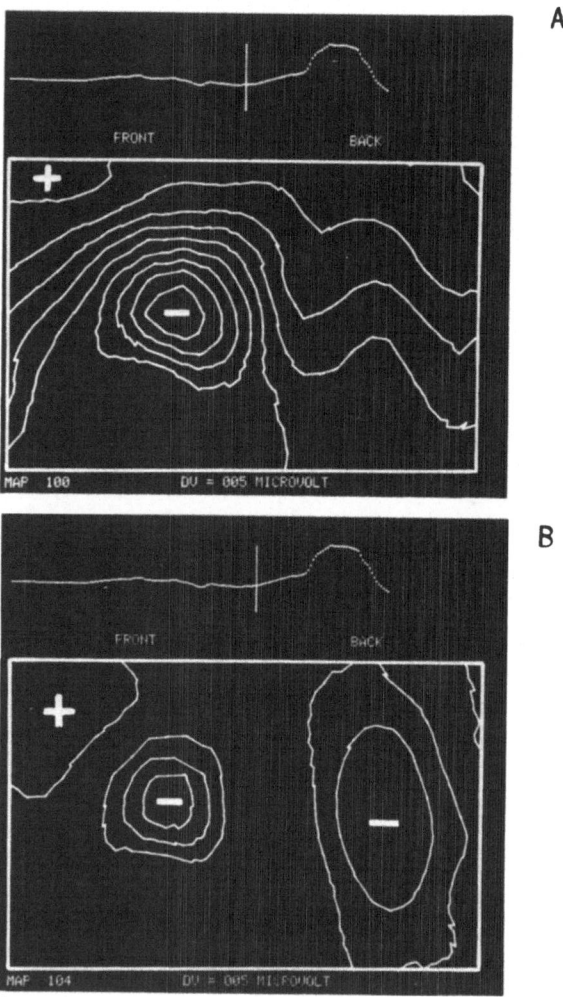

Fig. 3A and 3B (3C and 3D see next page)
     Surface potential maps recorded in the WPW patient at the
     instants indicated by the vertical bar on the electrocardio-
     gram. The increment between equipotential lines is 5 micro-
     volts in maps A,B,C and 20 microvolts in map D.
     A: potential patterns of atrial recovery;
     B,C: complicated patterns due to the overlapping of atrial
     recovery and initial ventricular pre-excitation potentials
     after 8 (B) and 12 (C) msec from A.
     D: pre-excitation potentials after 26 msec from A.

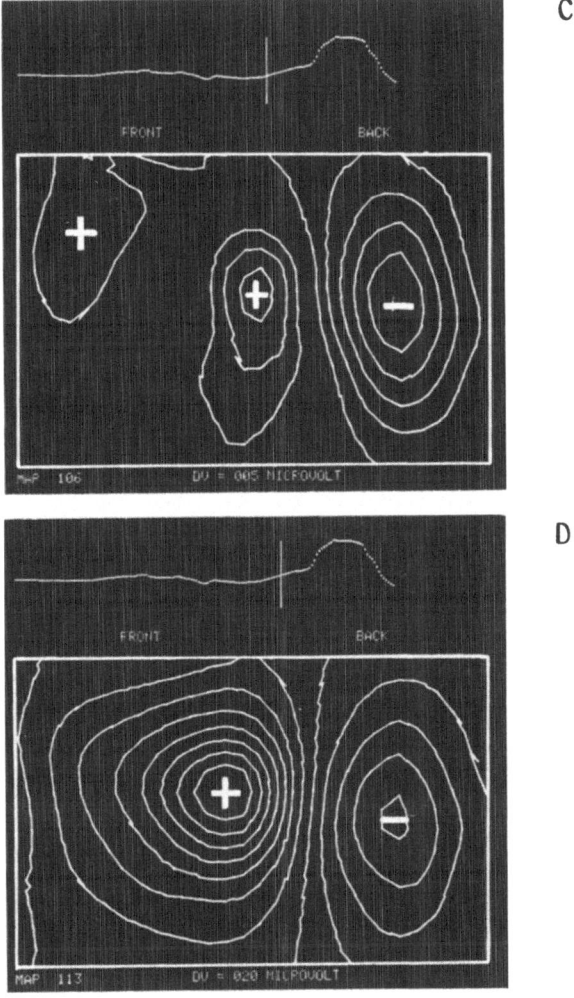

Fig. 3C and 3D   (Text see preceding page).

In our case it was $n_F = 19$, $n_T = 241$, $n_L = 163$, $n_H = 130$. The linear systems for the computation of the transfer matrices A, B, C were solved by means of the Jacobi iterative method and the convergence was assured by multiple deflations of Wielandt type. The calculations were carried out on a PDP 11/40 minicomputer.

RESULTS

Surface potentials recorded in the WPW patient revealed normal potential patterns during atrial excitation and the early stages of atrial recovery. The onset of pre-excitation revealed complicated patterns due to the overlapping of atrial recovery and ventricular pre-excitation potentials. During subsequent instants the pre-excitation potentials prevailed, giving rise to a single potential maximum on the left mammary region and a single potential minimum on the back (Fig. 3).

The search for the optimal dipole started 15 msec before the onset of pre-excitation and continued for another 40 msec. No physical meaning can be assigned to the optimal dipole elicited by the mathematical procedure before the onset of ventricular pre-excitation. At the beginning of pre-excitation the MRE showed a peak value for all locations (Fig. 4). This is probably due to the fact that the surface potential pattern is due to the simultaneous presence of two groups of sources in the heart: atrial recovery and ventricular pre-excitation. A single dipole cannot account for such a potential field. In the subsequent instants the ventricular source becomes preponderant and a single generator accounts reasonably well for the simple potential distribution. During this time interval location 14 in the posterior portion of the left A-V ring, close to the interventricular septum, generated the minimum MRE, while location 6 in the right lateral A-V ring produced the highest error (Fig. 5). All the other locations had intermediate MREs with increasing values appearing when we moved away from location 14 towards location 6.

The dipole moment for location 14 showed an abrupt change in direction at the time instant corresponding to the maximum value of the MRE (Fig. 4). During pre-excitation the dipole moment pointed anteriorly, leftward, with a small downward component, indicating the average direction of the advancing wavefront in the posterior ventricular wall.

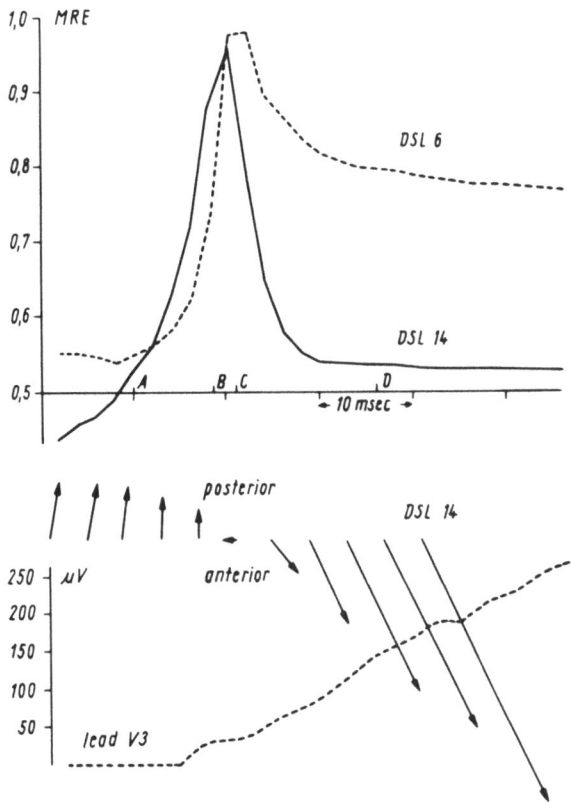

Fig. 4. At the top: MRE values before and after the onset of ventricular pre-excitation for the dipoles at locations 6 and 14
on the A-V rings.   A,B,C and D are the time instants referr-
ing to the surface potential maps of Fig. 3.
In the middle: horizontal projection of the optimal dipole
moment for location 14.
At the bottom: delta wave in lead V3 recorded in the WPW
patient.

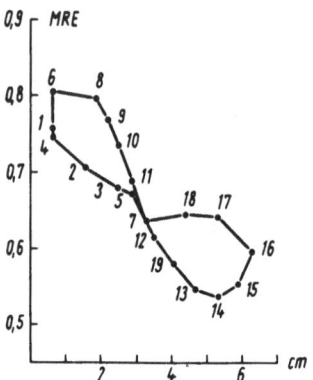

Fig. 5. MRE values for the 19 dipoles of the A-V rings at the time
       instant D of Figure 3 and 4.

DISCUSSION

     In the case study considered, the mathematical model localized
the pre-excitation site at a point on the left A-V ring, close to the
interventricular septum. This finding is in good agreement with independent results obtained from intracardiac recordings in the same
patient.

     The comparatively high value of the minimum MRE between recorded
and computed potential distributions for the optimal dipole might be
due to the intrinsic limitations of the model (errors in internal
geometry, conductivity, etc.), to the simultaneous presence of the
atrial repolarization sources throughout ventricular activation and
to the fact that a single dipole is a poor equivalent generator for
the pre-excitation wavefront, even in its early stage, as suggested
by recent data on the spread of excitation along and across muscle
fibers (Corbin and Scher, 1977; Baruffi et al., 1978). A more realistic equivalent generator might be taken into account after collect-
ing more data about the excitation wavefront and its associated current field.

     Further studies are necessary to confirm the validity of the
present model in different cases of WPW syndrome.

REFERENCES

Barnard, A.C.L., Duck, I.M., Lynn, M.S., 1967, The application of electromagnetic theory to electrocardiology, I: Derivation of the integral equations, Biophysical J., 7:443-462

Barnard, A.C.L., Duck, I.M., Lynn, M.S., and Timlake, W.P., 1967, The application of electromagnetic theory to electrocardiology. II: Numerical solution of the integral equations, Biophysical J., 7:463-491

Barnard, A.C.L., Holt, J.H., and Kramer, J.O., 1976, Models and methods in inverse electrocardiology: the UAB choice, in: "The theoretical basis of electrocardiology", C.V. Nelson and D.B. Geselowitz, eds., Clarendon Press, Oxford, 305-322

Barr, R.C., and Spach, M.S., 1976, Inverse solutions directly in terms of potentials, in: "The theoretical basis of electrocardiology", C.V. Nelson and D.B. Geselowitz, eds., Clarendon Press, Oxford, 294-304

Baruffi, S., Spaggiari, S., Stilli, D., Musso, E., and Taccardi, B., 1978, The importance of fiber orientation in determining the features of the cardiac electric field, in: "Modern electrocardiology", Z. Antalóczy, ed., Excerpta Medica, Amsterdam, 89-92

Colli Franzone, P., Gazzaniga, G., Guerri, L., Taccardi, B., and Viganotti, C., 1979, Accuracy evaluation in direct and inverse electrocardiology, in: "Progress in electrocardiology", P.W. Macfarlane, ed., Pitman Medical, Tunbridge Wells, 83-87

Corbin, L.V. II, and Scher, A.M., 1977, The canine heart as an electrocardiographic generator: dependence on cardiac cell orientation, Cir. Res., 41:58-67

De Ambroggi, L., Taccardi, B., and Macchi, E., 1976, Body surface maps of heart potentials. Tentative localization of pre-excited areas in forty-two Wolff-Parkinson-White patients, Circulation, 54:251-263

Eycleshymer, A.C., and Shoemaker, D.M., 1911, A cross section anatomy, Appleton-Century Crofts, New York

Savard, P., Roberge, F.A., Perry, J.B., and Nadeau, R.A., 1980, Representation of cardiac electrical activity by a moving dipole for normal and ectopic beats in the intant dog, Cir. Res., 46:415-425

Spach, M.S., Barr, R.C., and Lanning, C.F., 1978, Experimental basis for QRS and T wave potentials in the WPW syndrome. The relation of epicardial to body surface potential distributions in the intact chimpanzee, Cir. Res., 42:103-118

Yamada, K., et al., 1975, Body surface isopotential mapping in Wolff-Parkinson-White syndrome: non invasive methods to determine the localization of the accessory atrioventricular pathway, Am. Heart J., 90:721-734

LOCAL-INTEGRAL CHARACTERISTICS OF THE CARDIAC ELECTRICAL

GENERATOR BASED UPON THE MULTIPOLE EXPANSION OF THE POTENTIAL

P. Kneppo and L. I. Titomir

Institute of Measurement and Measuring
Technique of the Slovak Academy of Sciences
Bratislava, Czechoslovakia

The bioelectrical sources of the heart are distributed through the space of the myocardium or, in an approximate consideration of the depolarization process, over the surface of the excitation wave front. Thus, it is sometimes advantageous to use models or other equivalent generators of a distributed type for mathematical description of the cardiac electrical generator. They should be preferred in comparison with idealized point generators, the formulation and interpretation of which meets with known difficulties. So attention was recently paid to the theoretical possibility of representing the cardiac generator by current sources in the form of a double layer over a closed surface surrounding the heart (Geselowitz, 1976).

Let us suppose, that current sources are contained in a finite region of a homogeneous infinite volume conductor inside a closed surface S. Then it is possible to show that the potential outside the surface S is expressed as

$$\varphi = \frac{1}{4\pi} \int_S \varphi_s \, \text{grad} \left(\frac{1}{r} \right) \cdot d\overline{S} \tag{1}$$

where r is the distance from any point of the surface S to the point of measurement of the potential, $d\overline{S}$ is the vectorial element of the surface S, and φ_s is the potential which would exist on the surface S if a dielectric medium was outside S, i.e. the potential would be generated by the same sources situated on the surface of a homogeneous finite volume conductor bounded by the surface S.

On the other hand, a current double layer with the surface dipole moment density $\mathcal{J}$, distributed on a surface S in a homogeneous

61

infinite conductor with the resistivity ϱ, outside the surface S generates the potential

$$\varphi = \frac{\varrho}{4\pi} \int_S \mathcal{I} \operatorname{grad} \left(\frac{1}{\gamma} \right) \cdot d\overline{S} \qquad (2)$$

The comparison of the equations (1) and (2) shows, that if these two potentials are equal, i.e. the double layer is equivalent to the true sources, the following relation is valid:

$$\mathcal{I} = \frac{\varphi s}{\varrho} + \mathcal{I}_0 \qquad (3)$$

where $\mathcal{I}_0$ is an arbitrarily chosen term which is constant over the whole surface S. This term characterizes the surface dipole moment density of a uniform closed double layer which does not generate potentials in the external region with respect to the surface S.

Using the foregoing double layer as a description of the cardiac generator the surface S should be chosen as closely as possible to the true generators, so that the double layer most precisely reflects the structure of the true generator. For example, it may coincide with the epicardial surface of the heart ventricles or some other surfaces enveloping tightly enough the excitable muscle of the heart.

To define the function $\mathcal{I}$ it is necessary to find the first term of the equation (3) namely

$$\mathcal{I}_S = \frac{\varphi s}{\varrho}$$

and to choose the value $\mathcal{I}_0$. There are several approaches to determine $\mathcal{I}_S$ from measurements of the potential on the body surface and of coordinates of this surface (solution of the so-called inverse problem). In this paper we will consider the possibility to determine the function $\mathcal{I}_S$ approximately, using several initial members of a multipole expansion of the cardiac generator. Then it would be reasonable to choose such a value $\mathcal{I}_0$ for each time instant that the function $\mathcal{I}$ should be non-negative over the whole surface S. This means that the surface dipole moment density of the double layer would be directed from its internal to its external side in accordance with the predominantly radial direction of the propagation of excitation in the wall of the heart ventricles, as it is shown electrophysiologically.

Let the surface S be a sphere with the radius R, fully including the heart ventricles. Then it is possible to show that the main part of the surface dipole moment density

$$\mathcal{I}_S = \frac{\varphi s}{\varrho}$$

being variable over the surface of the double layer, may be expressed by the following series

$$\mathcal{J}_S = \frac{1}{4\pi} \sum_{n=1}^{\infty} \sum_{m=0}^{n} \frac{2n+1}{n} \cdot \frac{1}{R^{n+1}} (A_{nm} \cos m\psi + B_{nm} \sin m\psi) P_n^m (\cos \vartheta)$$

where A_{nm} and B_{nm} are the multipole components of the true cardiac generator, $P_n^m (\cos \vartheta)$ are the associated Legendre functions and ϑ, ψ are the angular co-ordinates of the point on the surface S in the spherical frame of reference with origin coinciding with the center of the sphere S.

In an actual electrocardiological investigation the multipole components of several lower orders can be calculated with reasonable accuracy using measured distributions of the potential on the external body surface and geometrical coordinates of this surface, if the body is assumed to be a homogeneous bounded conductor (Geselowitz, 1960).

For theoretical analysis we used in this work a multiple-layer spherical model including the regions of ventricular myocardium, intracavitary blood, lungs, external muscle layer of the thorax, and pericardium (Fig. 1). Each region has its own resistivity corresponding to known experimental data. Moreover, an estimation of the myocardial anisotropy was received using a simple two-layer model.

First of all we consider a small but finite element of the excitation wave propagating through the wall of the ventricles. This element may be represented by means of a point dipole with corresponding tangential and radial components (see the small circle in Fig. 1). It is necessary to find the contribution of such an elementary generator to the function $\mathcal{J}$, which is the surface dipole moment density of the equivalent double layer on the surface S. It is assumed that we may use only the multipole components of the first,

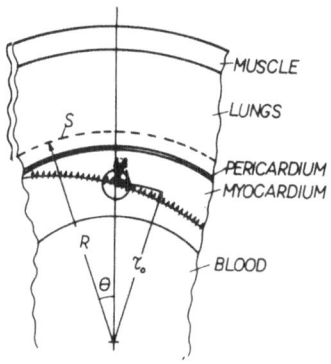

Fig. 1.

second and third orders. These components, namely A_{nm} and B_{nm} (n = 1, 2, 3) are found from the body surface potentials calculated for the homogeneous and inhomogeneous multiple-layer spherical models. In both cases the multipole components are determined assuming that the volume conductor is homogeneous. This means that in the second case the multipole components are distored in accordance with the inhomogeneity of the model. As the equations used for the calculation are rather cumbersome, although straightforward, they are not presented here. Then the approximate function $\mathcal{J}$ is calculated from these multipole components using equation (4), in which we take into account only the terms for n = 1, 2, 3.

The calculations are carried out separately for the tangential and radial dipoles, having equal dipole moments. The comparison of results shows that in the model with inhomogeneity of the body and anisotropy of the myocardium the function $\mathcal{J}$ has much smaller values for the tangential than for the radial dipole. Furthermore, in accordance with electrophysiological experimental data the tangential component of the elementary generators during the period of ventricular depolarization usually shows much smaller values in comparison with the radial component. Thus, the function $\mathcal{J}$ characterizes almost exclusively the radial component of the elementary generators.

Here we will present some results of the analysis for the radial component of the elementary generator. Figure 2 shows the normalized function $\mathcal{J}$ along a meridian of the sphere S for a radial dipole localized at the middle of the heart wall on the polar axis of the sphere (at $\theta = 0$). The thick solid line depicts the exact function $\mathcal{J}$ for the homogeneous model, the thin solid line the approximate function $\mathcal{J}$ (based upon the dipole, quadrupole and octupole components of the generator) also for the homogeneous model, and the dashed line the approximate function for the inhomogeneous model. These calculations were made for the surface S, coinciding with the epicardial surface of the heart ventricles (Fig. 2a) and for the surface S with a slightly greater radius, satisfying the requirements for equal contributions of dipole and quadrupole components into the maximal value of the function $\mathcal{J}$ (Fig. 2b). The results of the calculations provide the following conclusions. In all cases the function $\mathcal{J}$ has an explicit peak above the point of the localization of the generator. When using the lower-order multipole components the peak has a less acute shape, and additional extrema appear. The body inhomogeneity does not have a significant influence on the amplitude and the shape of the function $\mathcal{J}$. Small changes of the radius of the surface S have little effect on the function $\mathcal{J}$ too.

Figures 3a and 3b show the dependence of the maximal value of the function $\mathcal{J}$ upon the position of the radial dipole along the direction of the heart wall thickness. Changes of this position may correspond, for example, to a movement of the excitation wave from endocardium to epicardium of the ventricles. The notation used in Fig. 3 is the same as in Fig. 2.

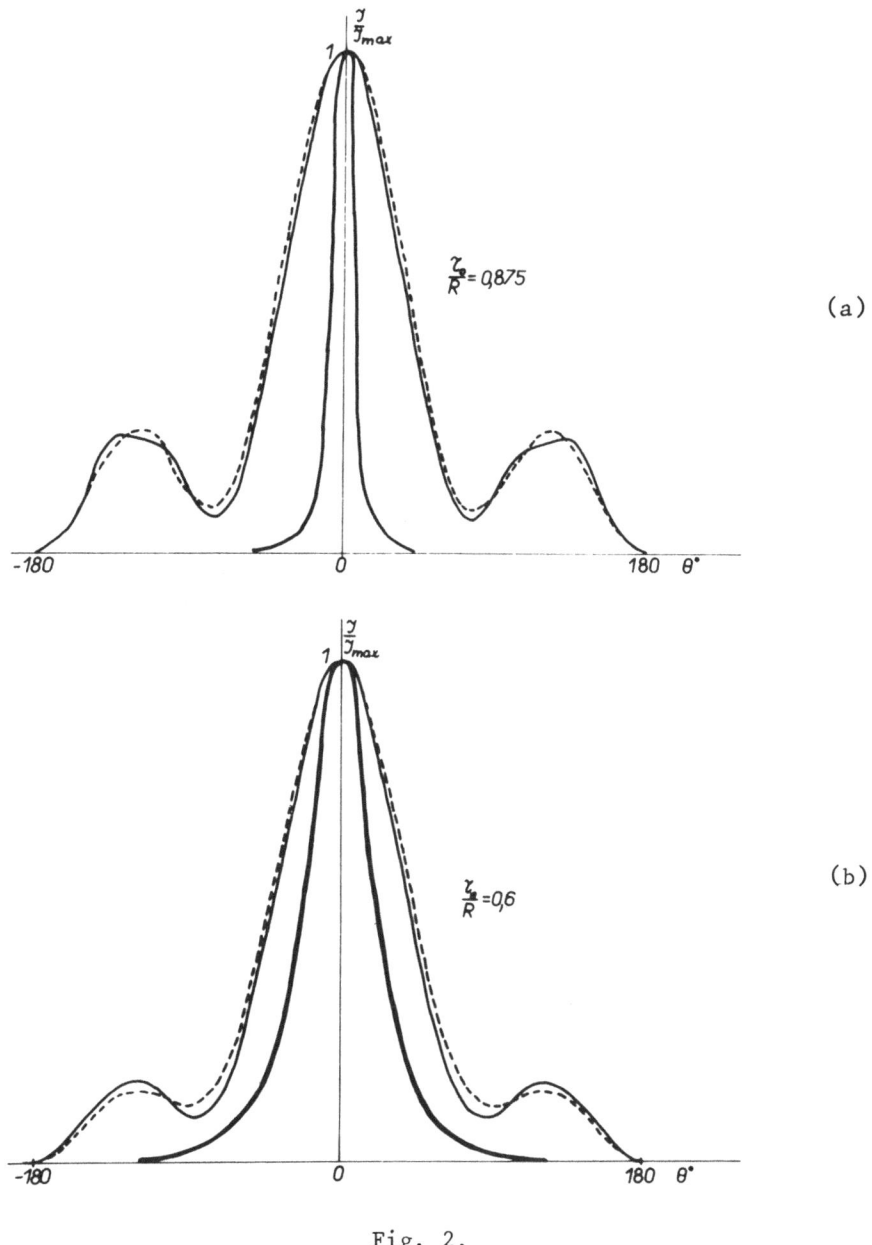

Fig. 2.

The graphs show that for the inhomogeneous volume conductor the maximal value of the function $\mathcal{J}$ is almost independent upon radial localization of the elementary generator. It means that all radial

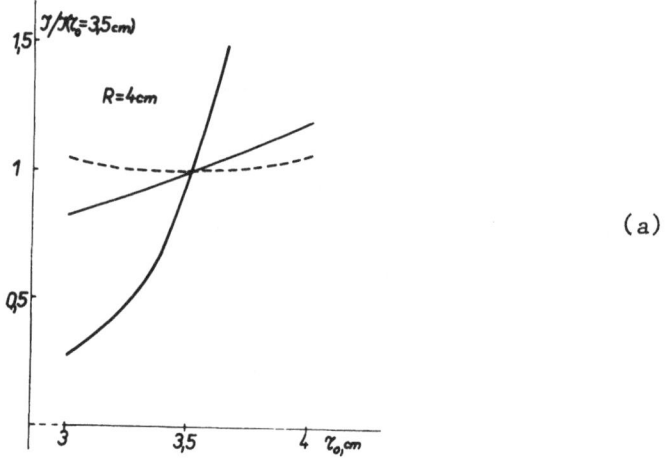

(a)

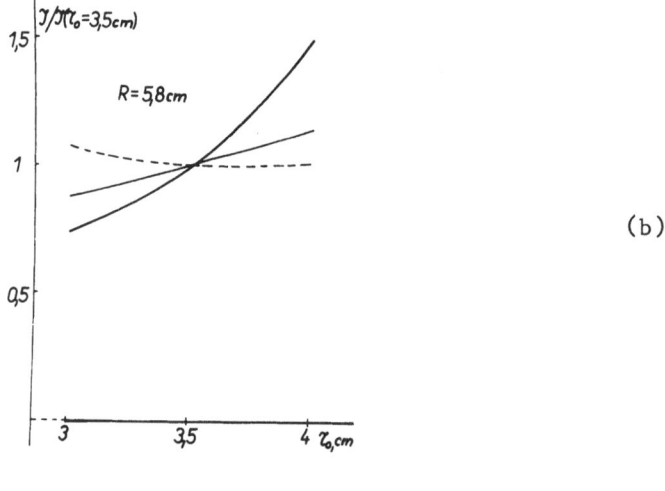

(b)

Fig. 3.

dipole generators localized on the same radial direction, but with different distances from the heart center, contribute to the function $\mathcal{J}$ with equal weights. If several such dipole generators or a distributed generator are present simultaneously along the same radial direction, the amplitude of the function $\mathcal{J}$ will be proportional to their sum or to the integral of the distributed generator.

Thus the function $\mathcal{J}$ distributed on the surface S has a peak indicating the angular localization of the projection of the dipole generator on this surface. At the same time the function $\mathcal{J}$ has rather weak sensitivity to the radial positions of the generators,

reflecting only their sum along radial directions. Therefore the
function $\mathcal{J}$ may be called a local-integral characteristic of the
cardiac electrical generator.

However, the actual excitation wave in the heart is a distributed current generator which may be approximated with sufficient
accuracy by a uniform double layer. Thus it is necessary also to
know how such distributed generator will be reflected by the function
$\mathcal{J}$.

We consider a model excitation wave having the form of a spherical segment of a uniform double layer, its axis of symmetry coinciding with the polar axis of the above mentioned spherical model of the
heart and the body (Fig. 4.). The spatial size of the excitation
wave, its diameter and angular sweep, is defined by the angle α. The
results of calculation of the function for this model of the excitation wave and for a homogeneous model of the body are presented in
Figs. 5a and 5b. Figure 5a shows the function $\mathcal{J}$ divided by the constant coefficient

$$K = \frac{M}{4} \left(\frac{r_0}{R}\right)^2 \ ,$$

where M is the constant value of the surface dipole moment density
of the model generator. The function $\mathcal{J}$ is calculated for several
values of the angle α using only the dipole, quadrupole and octupole
components of the model generator, as well as in the previous analysis. It is to be seen that in this case function $\mathcal{J}$ also has a more
or less explicit maximum, the angular position of which corresponds
to the localization of the excitation wave center. The amplitude and
shape of the main peak of the function depend upon the spatial size
of the excitation wave, i.e. upon the value of the angle α. For
$\alpha = 0°$ through $\alpha = 45°$ an increasing size of the wave leads to an
increase of both the amplitude and the width of the peak. However,
for greater values of α the amplitude of the peak increases significantly, and the width of the peak becomes the major feature reflecting the size of the excitation wave. This is illustrated in

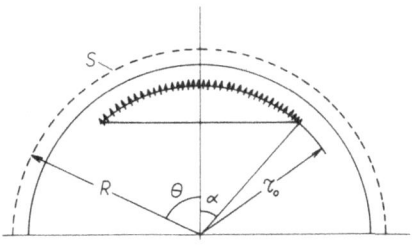

Fig. 4.

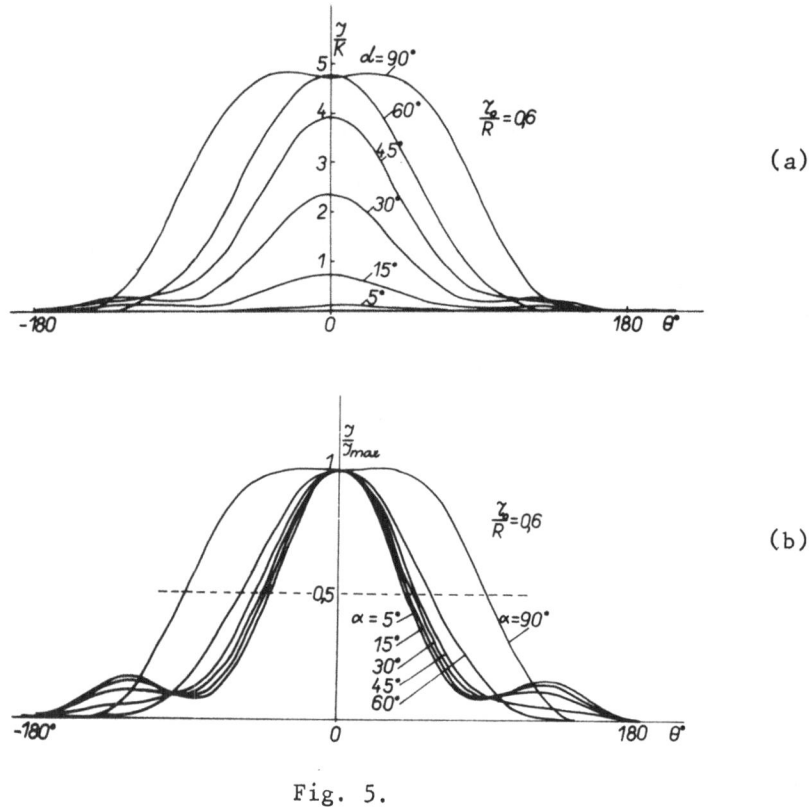

Fig. 5.

Fig. 5b, where the function γ for each α is divided by its maximal
value. It is to be seen that for α equal or greater than 45º the
width of the peak at the half-amplitude level is close to the angular
size of the excitation wave.

If there exist several simultaneously localized excitation waves
remotely enough from each other then the function would have the corresponding number of peaks, which characterize localization and size
of the waves. The resolving power of the local-integral characteristic γ will depend upon the number of multipole components used for
its formulation.

At present experiments are carried out to evaluate possibilities
of using the local-integral characteristics for analysing the electrical state of the human heart.

REFERENCES

Geselowitz, D. B., 1960, Multipole representation for an equivalent
        cardiac generator, Proc. IRE, 48: 75-79.
Geselowitz, D. B., 1976, Determination of multiple components, in:
        "The theoretical basis of electrocardiology," C.V.Nelson and
        D.B.Geselowitz, eds., Clarendon Press, Oxford, pp. 202-212.

A MATHEMATICAL MODEL OF THE EQUIVALENT MAGNETIC

CARDIAC GENERATOR

Pavel Tekel and Peter Kneppo

Institute of Measurement and Measuring
Technique of the Slovak Academy of Sciences
Bratislava, Czechoslovakia

The last ten years are noticed to be of increased interest in the study of the human heart as the source of an electromagnetic field that is periodically changing in time. For the complex investigation of the electromagnetic heart action it is necessary to investigate all components of electromagnetic field, i.e. the electric component and magnetic component, both being a function of space and time. While the electric component is studied and investigated for decades in different branches of electrocardiography, the magnetic component of the field is more systematically observed and studied only for several years. This was the consequence of the relatively delayed development of methods for measuring the extreme weak magnetic fields, to which the fields of biomagnetic origin belong.

The character of the electromagnetic cardiac field is known to be relatively complicated. It is necessary in studying the heart as the source of this field to create suitable, simplified and illustrative interpretations of this complicated source in the form of appropriate models of the heart - physical or mathematical ones.

This work deals with the analysis of the magnetic heart field based on the assumption of multipole character of the equivalent magnetic generator. The solution of the problem was based upon solving the direct task for the quasistationary case in magnetocardiography. When the presumption of the multipole character of the equivalent magnetic generator was accepted, then the inverse task was solved for the conditions of a magnetically homogeneous environment characterized by the permeability μ_0. The distribution of the scalar magnetic potential φ_M can be written as

$$\phi_M = \frac{1}{4\pi} \sum_{n=0}^{\infty} \sum_{m=0}^{n} \left[\frac{A_{nm}}{R^{n+1}} P_n^m (\cos\vartheta) \cos m\phi + \frac{B_{nm}}{R^{n+1}} P_n^m (\cos\vartheta) \sin m\phi \right] \quad (1)$$

It follows out of the solution of the direct task in the chosen spherical coordinate system (r_1, φ, θ) from the multipole magnetic source, which is characterized by the multipole coefficients A_{nm}, B_{nm} on the arbitrary spherical surface S. This surrounds the source with a radius R and with the center in the origin of the coordinate system.

The solving of the inverse problem consists in the determination of the multipole coefficients A_{nm}, B_{nm} appearing in this equation on the base of the known distribution of the scalar magnetic potential φ_M. The scalar magnetic potential φ_M is an auxiliary and immeasurable quantity. In practice it is necessary to determine the multipole coefficients A_{nm}, B_{nm} out of the value of a measurable quantity. The magnetic induction $\vec{B}$ is for instance of this character. The magnetic induction in the arbitrary point of our space outside the source is given by

$$\vec{B}(r_1, \phi, \vartheta) = \sum_{n=0}^{\infty} \sum_{m=0}^{n} \frac{\mu_0}{4\pi r_1^{n+2}} \{(n+1)P_n^m(\cos\vartheta)(A_{nm}\cos m\phi + B_{nm}\sin m\phi)\vec{e}_{r_1} +$$

$$\frac{m}{\sin\theta}P_n^m(\cos\vartheta)(A_{nm}\sin m\phi - B_{nm}\cos m\phi)\vec{e}_\phi$$

$$+ \left[P_n^{m+1}(\cos\vartheta) - m\,\text{ctg}\,\vartheta\ P_n^m(\cos\vartheta) \right] \cdot (A_{nm}\cos m\phi + B_{nm}\sin m\phi)\vec{e}_\vartheta\} \qquad (2)$$

$$= B_1\vec{e}_{r_1} + B_2\vec{e}_\phi + B_3\vec{e}_\vartheta$$

where $\vec{e}_{r_1}$, $\vec{e}_\varphi$, $\vec{e}_\vartheta$ are corresponding unit vectors in the spherical coordinate system.

It is possible to show that for an unambiguous determination of multipole coefficients the knowledge of the distribution of the radial component of the magnetic induction $\vec{B}_1$ on the mentioned spherical surface S is sufficient. We can write

$$\vec{B}_1(R, \phi, \theta) = \frac{\mu_0}{4\pi} \sum_{n=0}^{\infty} \sum_{m=0}^{n} \frac{1}{R^{n+2}} (n+1)P_n^m(\cos\vartheta)(A_{nm}\cos m\phi + B_{nm}\sin m\phi)\vec{e}_{r_1} \qquad (3)$$

Then the multipole coefficients A_{nm}, B_{nm} are given by

$$\begin{Bmatrix} A_{nm} \\ B_{nm} \end{Bmatrix} = \frac{1}{\mu_0} \frac{R^n}{n+1} \varepsilon_m \frac{(2n+1)(n-m)!}{(n+m)!} \oint_S B_1 P_n^m(\cos\vartheta) \begin{Bmatrix} \cos m\phi \\ \sin m\phi \end{Bmatrix} d,S \qquad (4)$$

where ε_m is s.c. Neumann's factor ($\varepsilon_m = 1$ if $m = 0$ and $\varepsilon_m = 2$ if $m \neq 0$). The previous expression of the coefficients A_{nm}, B_{nm} was gained for the case that the centre of the multipole is identical with the origin of the coordinate system. In this case we shift the magnetic multipole source from the point $(0,0,0)$ to the point $0'(x_o, y_o, z_o)$, being inside the spherical surface S, but assuming an unchanged distribution of magnetic induction on the spherical surface the multipole source will be characterized by new multipole coefficients A'_{nm}, B'_{nm}. They can be expressed unambiguously on the base of values of A_{nm}, B_{nm} and coordinates of shifting x_o, y_o, z_o. For instance the new coefficients for the dipole and quadrupole are given by

$$
\begin{aligned}
A'_{10} &= A_{10} \\
A'_{11} &= A_{11} \\
B'_{11} &= B_{11} \\
A'_{20} &= A_{20} - 2\,z_o A_{10} + x_o A_{11} + y_o B_{11} \\
A'_{21} &= A_{21} - x_o A_{10} - z_o A_{11} \\
A'_{22} &= A_{22} - 0{,}5\,x_o A_{11} + 0{,}5\,y_o B_{11} \\
B'_{21} &= B_{21} - z_o B_{11} - y_o A_{10} \\
B'_{22} &= B_{22} - 0{,}5\,y_o A_{11} - 0{,}5\,x_o B_{11}
\end{aligned}
\tag{5}
$$

The comparison of individual multipole contributions for producing the total field of the equivalent multipole generator was carried out in such a way, that the individual multipole contributions on the spherical surface S with radius R were compared to the total value of the mean quadratic scalar magnetic potential. The total mean quadratic scalar magnetic potential is defined by

$$
\phi_M^{RMS} = \sqrt{\frac{1}{S} \oint_S \phi_M^2 \, dS}
\tag{6}
$$

$$
\left[\phi_M^{RMS}\right]^2 = \sum_{n=0}^{\infty} \sum_{m=0}^{n} \left[(\phi_M^{RMS\,Anm})^2 + (\phi_M^{RMS\,Bnm})^2 \right]
\tag{7}
$$

$$
\left\{ \begin{matrix} \phi^{RMS\,Anm} \\ \phi^{RMS\,Bnm} \end{matrix} \right\} = \left\{ \begin{matrix} A_{nm} \\ B_{nm} \end{matrix} \right\} \sqrt{\frac{(n+m)!}{\varepsilon_m (2n+1)(n-m)!}}
\tag{8}
$$

On the assumption that the source of the magnetic field is sufficiently determined by the first two members of multipole expansion, i.e. by dipole and quadrupole, we may find such a point of localization $O'(x_0, y_0, z_0)$ of the equivalent generator, in which the contributions to the production of the equivalent magnetic generator field from the quadrupole will be a minimum and the whole equivalent generator will be in fact characterized by the dipole member of the expansion. Coordinates x_0, y_0, z_0 are calculated from a system of three equations gained on the base of the minimum of the function

$$\left[\phi_M^{RMS_{n=2}} \right]^2 = \frac{1}{16\pi^2 R^6} \sum_{m=0}^{2} \frac{(2+m)!}{5\varepsilon_m(2-m)!} (A_{nm}'^2 + B_{nm}'^2) \tag{9}$$

The ratio between the residual mean quadratic scalar magnetic potential from quadrupole and the mean quadratic scalar magnetic potential from dipole was shown to be the most advantageous criterion characterizing the contributions to the total potential of the multipole source from a multipole of the first and second order. This ratio is called complexity parameter K_Q:

$$K_Q = \frac{\phi_M^{RMS_{n=2}}}{\phi_M^{RMS_{n=1}}} = \sqrt{\frac{3}{5R^2} \cdot \frac{A_{20}'^2 + 3A_{21}'^2 + 12A_{22}'^2 + 3B_{21}'^2 + 12B_{22}'^2}{A_{10}^2 + A_{11}^2 + B_{11}^2}} \tag{10}$$

The more detailed discussion of the corresponding theory may be found in Tekel a.o. (1979).

The mentioned method enables to determine the equivalent magnetic generator in the form of a precisely determined magnetic dipole (orientation, magnitude, localisation) and the value of the complexity parameter for every moment of the heart action. Within one complete heart action it is possible to gain basic characteristics of the source. These are:

1) the course of the dipole moment components m_x, m_y, m_z
2) the trajectory of the s.c. magnetic center x_0, y_0, z_0 of the heart
3) the course of the complexity parameter K_Q.

Like in Kneppo a.o. (1979), these characteristics were named the magnetic integral characteristics of the heart.

The validity of the derived relations was tested on the mathematical model by using a digital computer. The simulation of the cardiac magnetic field was realized with the field of an exactly

defined magnetic dipole. The values of the radial component of magnetic induction of such a dipole were calculated on the spherical
surface S with the radius R = 10 cm in the points, which arose from
its angular division by 7,5°. These values of radial components of
magnetic induction represented the set of input data for the calculation of multipole coefficients of the dipole and quadrupole according
to the equation (4) as well as for the determination of the coordinates of the magnetic center of the heart x_o, y_o, z_o and the complexity parameter K_Q.

The aim of the experiment was to determine the exactness of the
determination of the dipole magnetic source which caused the given
distribution of radial components of magnetic induction in intensity,
magnitude and localization, according to the proposed algorithm. The
answer to this question may be gained by the comparison of the parameters of the input magnetic dipole, out of which the distribution of
radial components of magnetic induction was calculated, and parameters of dipole gained according to the described method. The software ensured the error calculation in the localization of the dipole
δ_r, in the determination of the dipole magnitude δ_m and in the determination of the orientation of the magnetic dipole moment ε. If we
denote the parameters of the input magnetic dipole

$$m_{xz}, \ m_{yz}, \ m_{zz}, \ |\vec{m}|_z = \sqrt{m^2_{xz} + m^2_{yz} + m^2_{zz}}$$

and its point of localization (x_{oz}, y_{oz}, z_{oz}) where the second index
"z" indicates "given" and parameters of the output dipole

$$m_x, \ m_y, \ m_z, \ |\vec{m}| = \sqrt{m^2_x + m^2_y + m^2_z},$$

then according to Fig. 1 the calculated errors are defined as follows

$$\delta_r[\%] = \frac{|\vec{\Delta r}|}{|\vec{r}_z|} \cdot 100 \tag{11}$$

$$\delta_m[\%] = \frac{|\vec{m}| - |\vec{m}|_z}{|\vec{m}|_z} \cdot 100 \tag{12}$$

$$\varepsilon = \frac{m_x m_{xz} + m_y m_{yz} + m_z m_{zz}}{\sqrt{m^2_x + m^2_y + m^2_z} \sqrt{m^2_{xz} + m^2_{yz} + m^2_{zz}}} \tag{13}$$

The influence of the separate input parameters on the exactness
of the determination of the final magnetic dipole was systematically
investigated and a set of 100 random selected cases of different input parameters was investigated too. In all cases the error in the
determination of the dipole localization was δ_r <6,5 %, the error

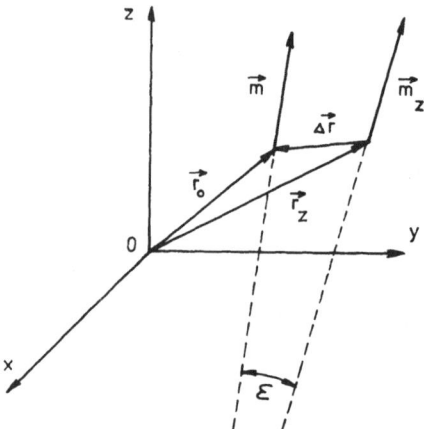

Fig. 1. Localization of the "input" and "output" magnetic dipoles:
"Input" dipole:

$$\vec{m}_z \ (m_{xz}, \ m_{yz}, \ m_{zz}), \ |\vec{m}|_z \ = \sqrt{m_{xz}^2 + m_{yz}^2 + m_{zz}^2}$$

"Output" dipole:

$$\vec{m} \ (m_x, \ m_y, \ m_z), \ |\vec{m}| \ = \sqrt{m_x^2 + m_y^2 + m_z^2}$$

in the determination of the dipole magnitude was δ_m <1,5 % and the
error in the determination of the orientation of the magnetic dipole
moment was ε <0,4°, when the shift of the input dipole from the
origin of the coordinate system was in the range 0 ÷ 0,7 R. While
making the computer experiments we were engaged also in the detection
of the influence of higher orders multipoles (quadrupole, octapole)
on the creation of the total magnetic field of an equivalent multi
pole generator in such a way, that the values of the adequate multipole contributions to the value of the total mean quadratic scalar
magnetic potential were determined. These contributions were determined in the case of multipole expansion in the origin of the coordinate system and the point of the magnetic center of the heart.
For the limitation of the structure of multipole generators in the
form of dipole + quadrupole + octapole the values of the mean quadratic scalar magnetic potentials for dipole, quadrupole and octapole
were marked as SMP02, SMP04 and SMP08, if the expansion was performed
in the origin and as SMP2, SMP4, SMP8, if the expansion was carried
out in the magnetic center of the heart. Experiments showed that the
typical course of this separate contribution is illustrated in Fig.
2. It is obvious, that the influence of quadrupole and octapole will
increase if the input data for the values of radial components of
magnetic induction are not of true dipole character. We may assume

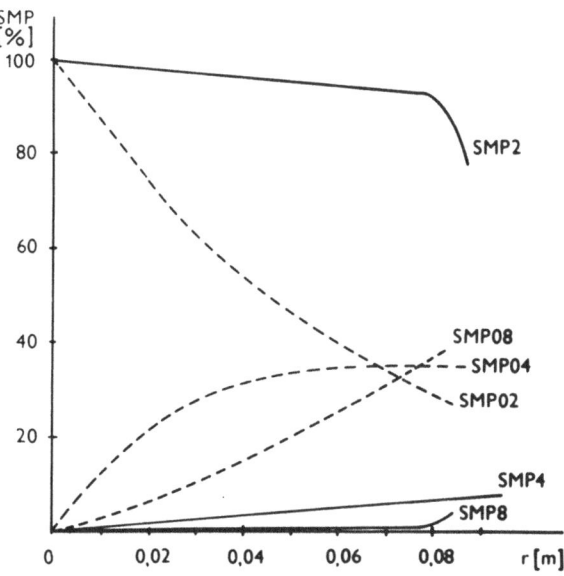

Fig. 2. Typical course of separate contributions to the total mean quadratic scalar magnetic potential of the equivalent magnetic generator in the form dipole + quadrupole + octopole.

that it will reach a value of about 10% of the total one. This indicates, that if we want to evaluate in the future the human magnetic cardiac field, which is not 100% of dipole character sufficiently precisely by the described method it will be necessary to take into account next to the dipole member also quadrupole or octapole members.

REFERENCES

Kneppo, P., Titomir, L. I., 1979, Integral characteristics of the human cardiac electrical generator from electric field measurements by means of an automatic cylindrical coordinator, IEEE Transactions on Biomedical Engineering, 26:21.

Tekel, P., and Kneppo, P., 1979, Integral characteristics of an equivalent magnetic generator of the heart, in: "Progress in Electrocardiology," P. W. Macfarlane, ed., Pitmann Medical, Tunbridge Wells, pp.146-150.

FACTORS INFLUENCING THE ST-SEGMENT MAPPING EVALUATION

(A MODEL STUDY)

Z. Drška

Institute of Physiological Regulations
Czechoslovak Academy of Sciences
Prague, Czechoslovakia

During recent years experiments were described trying to utilise ST isopotential praecordial maps for the evaluation of acute myocardial infarction and its development (Kronenberg et al., 1976; Madias and Wood, 1977; Hardarson et al., 1978; Murray et al., 1979).

From our previous papers follows that the surface potential distribution originating within the electrical field of the heart is markedly influenced by interactions of the excentrical position of the source of the electrical field of the heart, its spatial orientation, and the configuration of the thoracic surface (Drska et al., 1977).

The use of praecordial maps for the evaluation of the extent of ischaemia necessarily supposes that:

a) The moment (spatial magnitude) of the source equivalent to the region of ischaemia (in the given case of zone of ischaemia, surrounding the necrotic area) is proportional to the extent of the ischaemic or necrotic area.

b) The difference in spatial orientation of the above mentioned source does not significantly influence the ST potential sum in the praecordium during the development or regression of ischaemic area.

On the basis of vectorcardiographic experience especially it may be taken as proved that not only the magnitude of the equivalent source moment does change expressively, but also the spatial orientation of instantaneous vectors during the ST interval. This may be expected even in the case of a positive strain test when only changes are present in the extent and degree of ischaemia.

This paper attempts to verify to what extent the praecordial maps, in particular with the present mode of their evaluation, i.e. by the way of summing up potential values, meet the requirement for the purpose mentioned above (Jobin et al., 1976; Norris et al., 1976; Selwyn et al., 1977; Maser, 1978).

A current dipole was placed into an electrolytical tank-chest at the site corresponding to the position of the equivalent source in the actual electrical field of the heart (Frank, Seiden). The moment of the dipole was constant throughout all the measurements of this study.

Every setting of the dipolar axis was followed by measuring the potential distribution on the surface of the tank-chest in a regular network of points. Then the measured potentials were processed in a form of maps of isopotential lines (see Fig. 1), and, in the praecordial region the sum of the potential values was calculated.

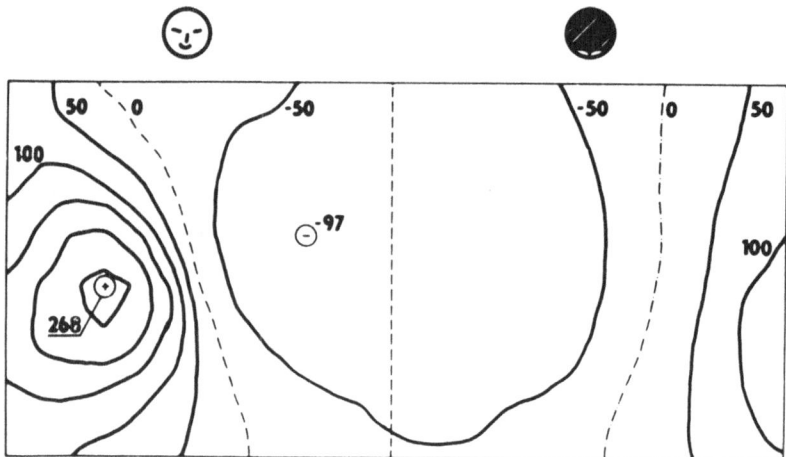

Fig. 1.   An example of a map of isopotential lines representing the potential distribution on the torso-model surface used in this study for determination of the praecordial ST segment potential summation.
Dipole direction: A = -160, E = 12

On the basis of vectorcardiographic data (azimuth and angle of elevation of the instantaneous vector at the beginning, in the centre and at the end of the ST interval) the axis of the dipole was successively set into the respective directions.

The vectorcardiographic data used were obtained from groups of persons after myocardial infarction in anteroseptal, anterior, lateral and posterior regions. The localization of the ischaemic area in the vectorcardiographic picture was verified by the strain test and by angiocardiography.

The results show an expressive dependence of the basic properties of praecordial maps on the spatial orientation of the source of the electric field (see Table 1). For instance, the change of the direction of the dipolar axis in space by 79° with changed azimuth from 45° to 116° and angle of elevation from -6° to -39°, leads to an expressive change of the sum of potential values in the praecordium from 200 to 89 mV. This occurs in spite of the fact that this difference of spatial directions does not exceed the variability of the direction of the instantaneous vector under investigation of one of the clinically differentiated localizations in myocardial infarction or local ischaemia.

A dipole was used in the model as an equivalent source. If a source, which is equivalent to the electrical field of the heart, possesses the properties of a source of higher order, i.e. quadrupole or octapole, the validity of the above mentioned relationships may be deduced even for such a case.

Table 1. The change of the potential sum in the praecordium of the torso-model during the simulation of the electrical field of the heart.
Value differences demonstrate the result of
the change of instantaneous vector direction
in the centre of ST interval.
A, E - azimuth and angle of elevation of the
dipole axis in degrees, ϕ - angle between
the dipole axes in space, ΣV - potential sum
divided by the number of derivations under
investigation in the praecordium of the
torso-model

A	E	ΣV	ϕ
45	- 6	200	
			79
116	-39	89	
			63
89	-19	141	

Under the presumption of an independent change of single components of this source (i.e. individual partial dipoles) on the basis of the superposition principle an increase of complexity of the relationships between the properties of the central or zonal ischaemia on one hand and between the properties of the potential distribution in the praecordium, in the given case the sum of potentials, within the ST interval on the other hand must be expected.

Following from the results of this study a clinically significant relationship between the characteristics of the praecordial maps and the area of zonal ischaemia and necrotic area on one side, or between their development in the given subject on the other side cannot be expected with the present way of evaluation.

REFERENCES

Drška, Z., Svoboda, P., and Novák, V., 1977, Maps of potential peak courses corresponding to vectorcardiograms. A model study, "Adv. Cardiol.," 19:84-87, Karger, Basel

Hardarson, T., Henning, H., Orourke, R.A., Karliner, J.S., Ryan, W., and Ross, J., 1978, Variability, reproducibility, and applications of precordial ST-segment mapping following acute myocardial infarction, Circulation, Vol. 57, No. 6

Jobin, G., Ostlund, J., Stankus, K., Chatterjee, K., Forrester, J.S., and Swan, H.J., 1976, Automated precordial mapping for S-T segment analysis: Variability in serial mappings of normal subjects and patients with stable S-T elevation, Am. J. Cardiol., 37:1052

Kronenberg, M.W., Hodges, M., Akiyama, T., Roberts, D.L., Ehrich, D.A., Biddle, T.L., and Yu, P.N., 1976, St-segment variations after acute myocardial infarction: Relation to clinical status, Circulation, 54:756

Madias, J.E., and Wood, W.B. Jr., 1977, Precordial ST-segment mapping. 3. Stability of maps in the early phase of acute myocardial infarction, A. Heart J., 93:603-609

Maseri, A., 1978, Coronary artery spasm - diagnostic and therapeutic implications, A. Heart J., 96, No. 4

Murray, R.G., Peshock, R.M., Parkey, R.W., Bonte, F.J., Willerson, J.T., and Blomqvist, C.G., 1979, ST isopotential precordial surface maps in patients with acute myocardial infarction, J. Electrocardiology 12, 1:55-64

Norris, R.M., Barratt-Boyes, C., Heng, M.K., and Singh, B.N., 1976, Failure of ST segment elevation to predict severity of acute myocardial infarction, Brit. Heart J., 38:85-92

Selwyn, A.P., Ogunro, E.A., and Shillingford, J.P., 1977, Natural history and evaluation of ST segment changes and MB CK release in acute myocardial infarction, Brit. Heart J., 39:988-994

MODEL STUDY OF THE LEAD SYSTEM INFLUENCE ON THE IMAGE

OF THE PHYSICAL EQUIVALENT GENERATOR

Vavrinec Szathmáry

Institute of Normal and Pathological Physiology
Centre of Physiological Sciences, Slovak
Academy of Sciences, Bratislava, Czechoslovakia

A number of investigators have studied the quality of vector-cardiographic lead systems. For instance Macfarlane and Lawrie (1974) proposed the correction of potential values obtained by means of McFee-Parungao's corrected orthogonal lead system. The present study was undertaken to observe the reflection of the dynamic changes of an actual current source in its image, determined by means of either McFee-Parungao's or Frank's orthogonal lead system.

The model measurements were realized in an electrolytic tank which was shaped to fit the torso of a normal adult man. The fibreplast tank was filled up with tap water, which we used as a homogeneous conducting medium. The schematic diagram of the experimental setup is shown in Figure 1, where number 1 marks the function generator (Type TR-0452), number 2 the oscilloscope (OPD 280-U), number 3 the convenient lead system and number 4 the electrolytic tank. As the actual current source we used a physical dipole, created by two nickel spheres with an interpole distance of 3 cm. The dipole was energized by 100 Hz sinusoidal current. The measurements were realized for 7 different dipole positions in the tank, see Figure 2. There were 6 dipole localizations at the transections of the borderline of the heart region and the coordinate axes with origin at the point HC. This point represents the center of the heart region. The spatial localization of this point in the electrolytic tank is to be seen in Figure 3. The horizontal level, which includes this point, intersects the torso 3 cm above the lower border of the processus xyphoideus. In this plane the center of the heart region was placed at one third of the chest thickness and width respectively. The other dipole localizations were situated 4 cm away from the center HC. In each dipole position we gradually varied the azimuth in steps of 10° from 0° to 200°. The elevation was constant for all measure-

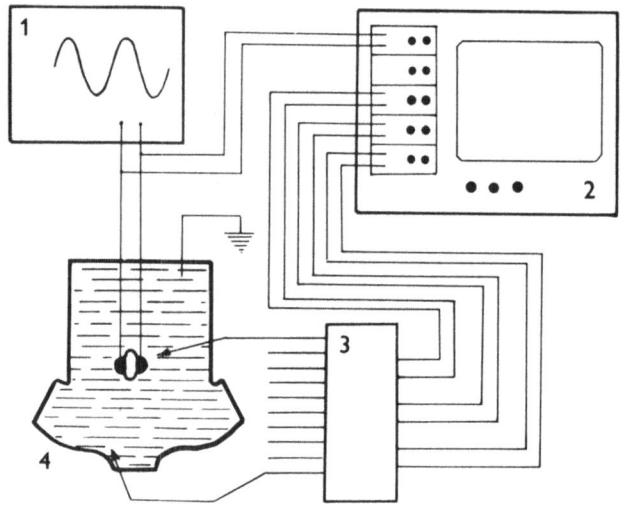

Fig. 1. Schematic diagram of the experimental setup

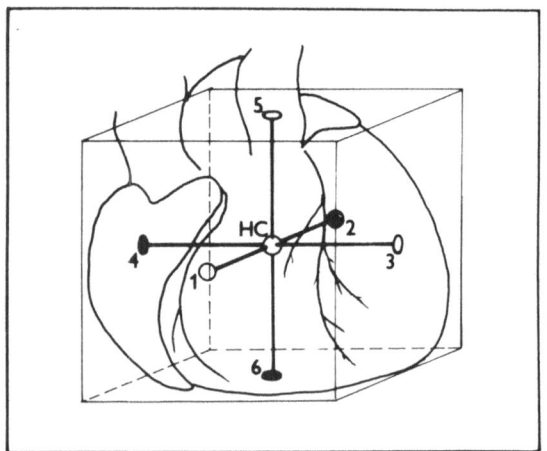

Fig. 2. The dipole positions around the heart centre HC

ments and equal to 0°. From the obtained x, y and z components of
the electric potentials we derived the parameters of the image
sources. Evaluating the results we were mainly interested in the
comparison of the changes in orientation of the actual source to the
same changes of the image source.

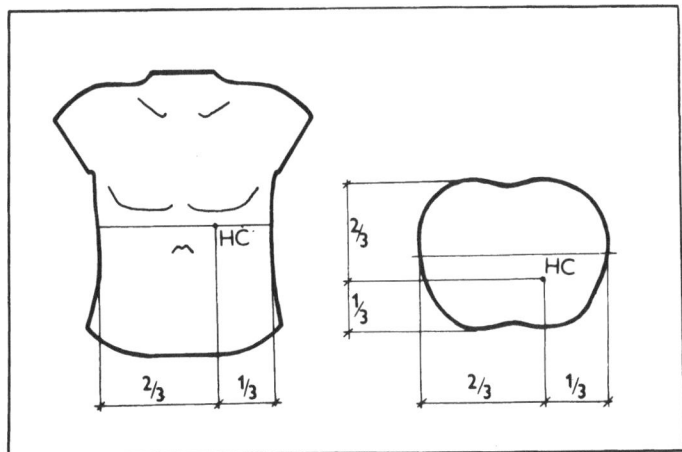

Fig. 3. The spatial localization of the heart center HC in the homogeneous electrolytic tank.

     The graphical illustration of the obtained results is shown in
Figure 4. The values below the abscissa express the fact, that
changes of the angle of the image dipole orientation were smaller,
and the values above the abscissa - that changes of the angle were
larger than the corresponding changes of the actual source. One unit
at the axes scale is related to the 10° steps of the changes in
azimuth and means either 1° or 10%. On the left part the results are
presented, which were obtained using the lead system of McFee-
Parungao, and on the right part those using the Frank's system. As
it can be seen the lead systems under investigation had a slowing
influence on the image of dipole rotation when the dipole was oriented anteriorly or posteriorly and an accelerating influence, when the
dipole had a lateral orientation. Only in the case, when the dipole
was localized at the extremely left position, the above mentioned
influences were opposite. This is through exception of only one
dipole position. The magnitudes of these angular speed differences
are proportionally related to the excentricity of the dipole localization in the electrolytic tank. There was also a difference caused
by the two lead systems used for investigation. The regions of the
slowing and accelerating effects of the Frank's lead system are
shifted in relation to those of the McFee's lead system with angles,
which depend on the dipole position. These results are in agreement
with theoretical studies, obtained by mathematical modelling, when
the torso was replaced by a homogeneous conducting sphere.

     For electrophysiological or clinical application and evaluation
of the above mentioned lead system differences and their influences
on the measured heart potentials two different facts have to be re-
spected. During the cardiac activation all parts of the myocardium

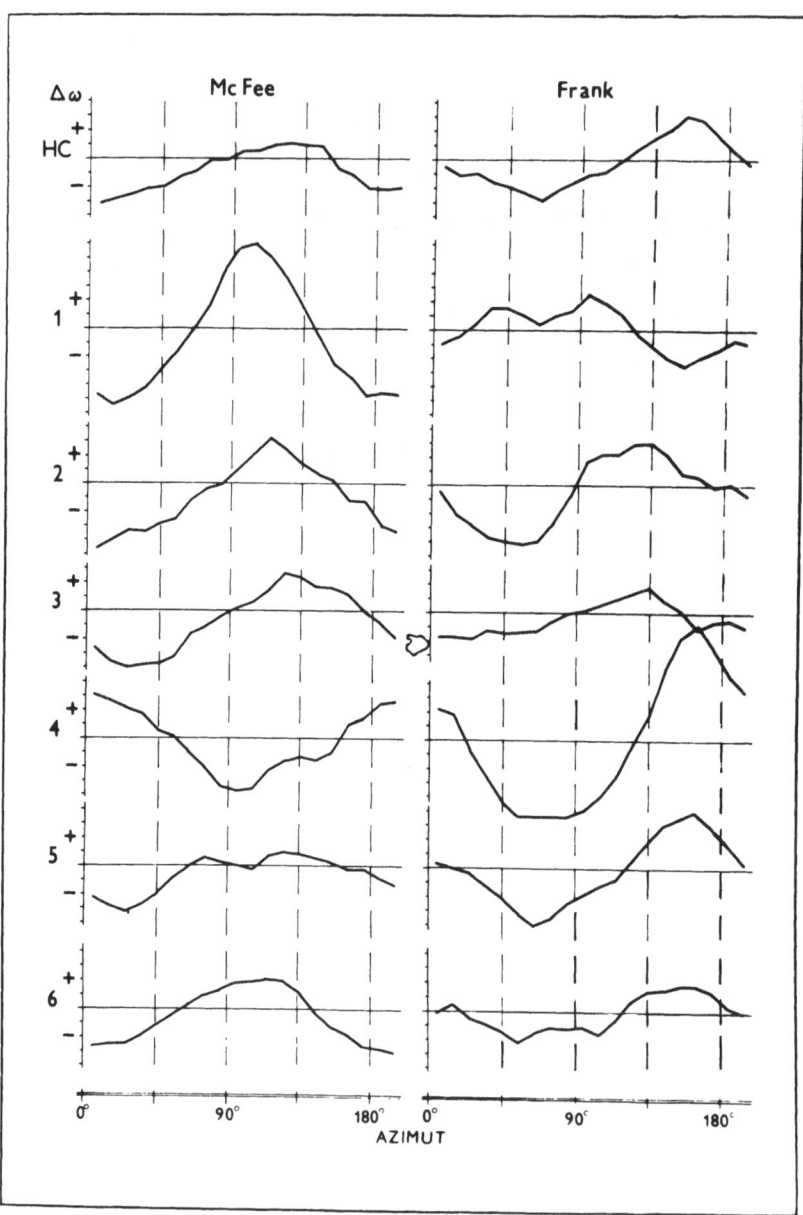

Fig. 4. The graphical illustration of the differences between the
changes of orientation of the actual dipole sources and
their images for the two compared lead systems.

are effective as sources of the electric field, and the localization of the resulting heart vector lies mainly at the middle part of the heart region.

REFERENCE

Macfarlane, P. W., Lawrie, T. D. V., 1974, An Introduction to Automated Electrocardiogram Interpretation - Computers in Medicine Series, D. W. Hill, Butterworth and Comp., General Ed., pp 115

3. MEASUREMENTS IN THE CARDIAC ELECTRIC FIELD

PROBLEMS AND TRENDS IN BODY SURFACE MAPPING OF

BIOELECTRIC FIELDS

Bruno Taccardi*

Istituto di Fisiologia Generale e
Centro Simes, Universitá di Parma
Parma, Italy

Bioelectric phenomena are generally studied by recording and displaying local potential changes as a function of time. Electrocardiography and electroencephalography are typical examples of this approach. On the other hand, a great deal of information on the topography and time-course of electrophysiological events in the heart, skeletal muscle and the nervous system has been recently obtained by another method, i.e. by mapping the instantaneous distribution of bioelectric potentials in two or three dimensions.

Application of this method to cardiac electrophysiology and clinical cardiology has given stimulating results. Measuring the potential field within the thickness of the ventricular walls, on the epicardium and the body surface has definitely improved our understanding of cardiac electrical events. Recent attempts to display the topographical distribution of evoked potentials on the scalp have also shed new light on sensory and motor mechanisms.

On the other hand, a widespread use of the mapping method is still hampered by a number of technical and theoretical problems relating to data acquisition, processing, classification and interpretation.

DATA ACQUISITION AND PROCESSING

The number of leads that must be recorded in order to define the distribution of heart potentials on the body surface with sufficient

*Supported by grant N. 79.01258.86 of the "Consiglio Nazionale delle Recerche, Progetto speciale di Tecnologie Biomediche, sottoprogetto Bioimmagini 2".

accuracy has not been established so far. Figures ranging between
9 and 256 have been proposed. The solution for this problem can be
found by measuring the spatial frequency contents in body surface
maps. A group of scientists is now working in this area at the
Institute for Applied Mathematics (I.A.C.) in Rome (Barone, Ciarlini,
Regoliosi, this Symposium). The visual inspection of maps suggests
that the distribution of spatial frequencies is not constant over the
surface of the chest, and also varies during the heart cycle. Once
the optimum number and location of leads will be established, it may
be useful to record more leads than are theoretically necessary, in
order to compensate the noise present in the signals. On the other
hand, if it is necessary to keep the number of leads as low as possible, as often occurs in clinical practice, transfer coefficients
can be used, which enable the whole map to be calculated from 20 to
30 leads, with comparatively small errors. This work has been recently developed by Barr et al., from the Duke University group in
Durham, N.C. and by Lux et al. in Salt Lake City.

Other problems arise when we try to choose the most appropriate
baseline for measuring instantaneous amplitudes in the electrocardiograms. Theoretically, the T-P interval should be chosen.
However, slight potential changes are often observed during that
interval too. These may be due to late U-wave effects, particularly
when the heart rate is high, or may depend on the weak potential
field that is generated by the ventricles when they are slowly filling with blood. On the other hand, atrial recovery potentials
often outlast the QRS interval and superimpose themselves to QRS and
early ST potentials. By choosing the P-Q level as a baseline, the
distorting effect of atrial recovery potentials may be partially eliminated but other errors are introduced. Baseline drifts are also
disturbing events, particularly when low signals are to be measured.
At present, interpolation procedures are available for correcting
not only linear, but also non linear drifts.

The choice of a correct baseline is theoretically impossible in
subjects with acute myocardial infarction or with atrial or ventricular fibrillation, since in those cases there is no time-interval
during which the heart does not generate a current field. Despite
this difficulty, chest maps provide useful information in acute myocardial infarction, as will be discussed in the following pages.
In these patients the magnetocardiogram may be of considerable help
in detecting "systolic" and "diastolic" injury potentials that cannot be measured properly by conventional electrocardiography.

Additional problems arise when we try to map the low potential
fields which exist during 65% of the heart cycle, namely during the
P, PQ, ST and U intervals. Here, the signal to noise ratio is so
low that the maps may become meaningless if constructed by the usual
methods, since they mostly depict the distribution of noise. This
consideration applies not only to human maps, but also to the low

signals obtained under other experimental conditions, e.g. when potentials are recorded from conducting volumes, surrounding isolated
hearts. Spatial filtering definitely improves the signal to noise
ratio and results in readable maps. Extensive studies in this area
have been carried out by the I.A.C. group in Rome. Spatial filtering
can be achieved by the "moving average" procedure, by selecting a
frequency domain in the bi-dimensional Fourier spectrum, or by
smoothing the original data with spline functions. Unfortunately
these procedures, when applied to the same data, do not yield identical results. Thus, much work remains to be done in order to evaluate the reliability of filtered maps.

     Another method for increasing the signal to noise ratio consists
in averaging signals relating to many heart beats. The procedure is
acceptable only if all the beats selected for averaging are electrically identical. This is difficult to ascertain, since part of the
signal is obscured by the noise; nevertheless, physiological and
statistical methods can be used to restrict the selection to quasi-
identical beats.

     The equipment for mapping bioelectric fields can be optimized
only after the problems discussed above have been solved. At present
two types of instruments are available, both in our own laboratory
and in other centers: large systems for research work and small instruments for clinical purposes. The large systems record up to 256
signals simultaneously and store them in digital form on tape or
disc. The recorded data are then processed by a mini-computer,
either on-line or off-line, and finally presented on a graphic display unit. These machines are usually bulky and expensive, but their
versatility is indispensable to explore all the possible developments
of the mapping method. More recently, highly automated, small-size
instruments have been built which enable the thoracic or epicardial
electric fields to be mapped on-line at the bedside in coronary care
units, operating rooms and hospital wards. Thus, useful information
on infarct size can be obtained by displaying, albeit with some dis-
tortion, the distribution of superimposed excitation injury and recovery potentials on the chest surface. In this way, the effectiveness of medical or surgical treatment can be evaluated on-line every
few minutes.

DATA BANKS AND CLASSIFICATION OF MAPS

     The lack of a complete library of normal and abnormal maps is
one of the obstacles that still prevent body surface mapping from
being widely used as a clinical tool. An ideal data bank should contain body surface maps relating to 100 or more subjects for each age
group, sex, weight and heart condition. Such a large bank is not yet
available, even for traditional electrocardiography. However, many
laboratories have collected maps in hundreds of normal and abnormal

cases, and a number of classification criteria are now available. The "Difference maps" and "Departure maps" which were proposed by Flowers, Horan et al. in 1976 are a promising tool for differentiating normal from abnormal subjects. Other discriminant parameters are now being studied by our group in Parma and by others, namely: the trajectory of maxima and minima, their time-course, the integral of the potential function over positive or negative areas at specific time instants, the time course of the potential difference between maximum and minimum etc.

The time-integral of the potential over the entire QRST interval (i.e. Wilson's "gradient") at every point on the surface of the chest has been recently measured by Abildskov and his co-workers, who displayed their results as "isointegral maps" with a view to detecting the tendency to arrhythmias, which is often associated with an irregular distribution of recovery times, in ventricular muscle. Other important parameters for revealing electrical abnormalities can be deduced from chest maps recorded during exercise tests. Such maps have been recorded by Schubert et al., Block et al., Fox et al. with interesting results.

DETERMINISTIC INTERPRETATION OF POTENTIAL DISTRIBUTIONS

Interpretation of potential fields in terms of intracardiac events can be attempted through "direct" and "inverse" approaches. Suitable algorithms are now available for solving the "direct" problem, i.e. calculating the potential field from known current generators, even in non-homogeneous conducting media of irregular shape, such as the human body. However, a reliable model for the generator itself is still lacking. For many years the excitation wave front has been considered to act as a uniform dipole layer. In 1976 Corbin and Scher produced evidence against this contention and stressed the importance of the fiber direction in determining the anisotropic nature of the current sources. Further models taking into consideration the differences in myocardial resistivity and conduction velocity were developed by M. Spach et al., A. van Oosterom, R. Plonsey and by others.

Studies from the Parma group have shown that a distribution of heart potentials in a cylindrical tank surrounding an isolated heart differs from that which would be expected from the widely accepted solid angle theory (Baruffi et al., this Symposium).

Due to the difficulty of defining the cardiac sources, most efforts to solve the inverse problem (i.e. inferring cardiac events from measurements taken at a distance) have been aimed at calculating potential distributions at the surface of the heart, a goal that can be achieved without making any assumption on the nature and geometry of intracardiac sources. This work has been done by the Duke University group, by the Pavia group (Colli Franzone et al., this Symposium) and by others. Serious difficulties have been encountered, due to the fact that small errors in potential measurements give rise to large errors in the calculated "epicardial" maps. These errors can be reduced by using regularizing procedures.

On the other hand, the position, strength and orientation of a single intracardiac dipole can be determined by using a combination of direct and inverse models, as will be shown during this symposium (Macchi et al., this Symposium).

Despite the difficulties mentioned on the preceding pages, bi- and tridimensional maps of heart potentials are now used in many countries as a research tool and also as a diagnostic method. The distribution of heart potentials in the ventricular walls and in conducting media surrounding isolated hearts has suggested that the classical models, which represent the excitation wavefront as a uniform dipole layer may be inadequate. Epicardial maps in experimental animals have revealed that the spread of excitation on the ventricular surface is much more complicated than was previously known. Electroencephalographic maps, depicting the distribution of evoked potentials on the scalp, yield new information on the topographic sequence of cortical or subcortical events after a peripheral stimulus. In the clinical practice, body surface maps have provided diagnostic information for acute and old myocardial infarctions, coronary insufficiency, right and left ventricular hypertrophy, complicated and uncomplicated conduction disturbances, arrhythmias etc. At surgery, epicardial maps have been used to detect the origin of intractable arrhythmias. It may be expected that the increasing availability of low-cost computers will promote a wider use of this method in research laboratories and medical centers.

A SYSTEM TO STANDARDIZE AND ANALYSE SURFACE

MEASUREMENTS FOR MODEL FITTING

P. Barone, P. Ciarlini, G. Regoliosi

Istituto per l'Applicazione del Calcolo, CNR
Roma, Italy

INTRODUCTION

The knowledge of the surface cardiac potentials is limited as far as the amount and accuracy of measures by the instrument features is regarded. In particular, for many applications, the information content of data recorded in a "standard" way can be insufficient, for instance: the quantitative studies of surface potentials in space-time intervals characterized by high gradients or low signal/noise ratio, or the inverse problem, where such data are inputs, or the direct problem, where they are used for control purposes, or the black-box surface modelling, etc.

To overcome such instrumental limitations a possible solution is based on the construction of software systems. If the data acquisition is controlled by a minicomputer, fitted with sophisticated software, the instrumentation capabilities can be increased, with a cost, measured in computing time, proportional to the desired sophistication of the results.

That is to say also that, with a given instrumentation, the betterment of cost/effectiveness can be achieved improving the data acquisition through sophistication of the software systems.

According to our analysis the main functions of the needed software are:

- to acquire potential measurements for any desired space-time resolution (within certain limits),
- to remove the effect of deterministic error and reduce the random one.

The first feature can be reached by synchronizing several recordings taken in different areas of the thorax (different topology) to reconstruct composite maps. The desired time resolution can be achieved using a multiplexer characterized by a constant ratio between the sampled data number per cycle and the cycle length. Moreover the software should provide data belonging to any space-time window. In some sense, this software should simulate an acquisitional system performing at a level as high as to be requested transparent.

The second feature can be reached through the following steps:
1. software calibration of the instrument, particularly of the amplifiers and the A/D converter, for removing the corresponding systematic errors;
2. estimate of the base-line and of the systematic error induced by the environment at the acquisition time (50-cycle noise, drifts, etc.) and subsequent subtraction;
3. reduction of the random error by "averaging" several homogeneous beats, when it is not possible to do it by hardware.

The data which are preprocessed according to the above stated procedure enjoy very useful statistical features; in particular they are unbiased and with an error which is independent and approximately normal.

To introduce the CPPO modular system which is working for the surface mapping laboratory at Parma University, some information on the instrumentation and on experiments conducted at this Center seems to be useful.

INSTRUMENTATION AND EXPERIMENTS

The Parma laboratory uses a 240-probe instrument, with a 2 msec cycle multiplexer (Cottini et al., 1972). The software system, named GISPEM implemented on a PDP11/40, enables to store, in real time, 120.000 data per second in 240-byte records (Barone et al., 1977).

Consequently an experiment is considered "standard" when it is characterized by 2 msec sampling time, a location of 240 probes in a fixed area of the thorax, in accordance with a fixed sampling, and finally by a single data set (in 240-value blocks).

A "non-standard" experiment is then characterized by a sampling time of 2^{1-n} msec; n=0, 1, 2, 3, 4, and/or any probe location. A data set is obtained, divided in $240/2^n$ value (logical) records, for each different portion of the chosen areas.

The Parma instrumentation includes also a dynamic buffer memory controlled by a microprocessor which may average several (homogeneous) beats in real time.

CPPO Structure

CPPO is a system composed by the following four modules:

CONFIL - which converts each data set in the proper format for the experiment and assigns the corresponding topology;

PREFIL - which eliminates systematic error from each data set;

PARALL - which synchronizes several data sets and chooses a space-time window, parallelepiped-shaped;

ORDINT - which divides each parallelepiped in one or two dimensional sections and interpolates possible missing data.

In order to graphically display such sections, the system is connected with two interactive systems for plotting plane curves (Ocello, 1979) and surfaces, known on a regular grid, on a Tektronix screen.

Each module is a single job and it is entry-point organized. The user can reach an entry-point at any time since the system checks if the request is admissible with respect to the executed functions and available information.

So each function can be executed many times with different parameter values.

The user-system interaction is accomplished through the display screen, which allows to input numerical or alphanumerical data and output numerical or graphical results (location of probes on the thorax, parameter values, error messages).

Modules communicate among each other through secondary memory, therefore the user works essentially as if the system was composed by a single job.

The modular structure was chosen, both for mini-computer limitation and for the necessity to distinguish between operations to be done on each data set and those on the whole experiment. Moreover other modules (for instance statistical analysis) are easily connectable to the system.

CPPO Modules

CONFIL - This module is an interface between GISPEM and the following modules, since it converts a data set from the fixed GISPEM format into the proper format for the experiment.

Moreover, since the potential measurements can be taken in different thoracic areas and with different spatial sampling length, the module allows the user to describe, in an interactive way, the actual spatial topology (with respect to a regular fixed grid) and as a second operation it stores all the information, characterizing the experiment.

PREFIL - The main function of this module is to remove the systematic error. It is supposed that amplifier thermal drifts, 50-cycle noise and other known periodic components, if not eliminated by other techniques, be well modelled by sinusoids of opportune frequency.

For each amplifier, error is modelled as follows:

$$f(t;\bar{\omega},\bar{c}) = c_o + \sum_1^f j \left[c_j \cos (2\pi\omega_j t) + c_{j+f} \sin (2\pi\omega_j t) \right]$$

where: c_0 represents the base-line level, f is the number of non-zero frequencies ω_j, which describe systematic components.

Since f and ω_j could not be exactly known a-priori, the least square estimate of $\bar{c}$ parameters can be computed several times until an adequate and significant model is obtained.

To get a correct estimate of the model, intervals, where the cardiac signal is zero, are to be used. These intervals can be evaluated in a graphical interactive way.

Of course PREFIL removes, previously to the model estimate, systematic errors caused by the A/D converter and amplifiers from data. Hence code level and amplifier gain estimates are a necessary equipment of each recorded set.

In the module the time reference value of the actual recording is estimated.

This reference, defined as the first instant where potential gets zero after R-wave, is evaluated, using least square methods, at the least resolution compatible with the acquisition instrument. The reference value will be used in the following module, to synchronize data sets.

PARALL - This module works on different data belonging to the same experiment. Two kinds of functions are implemented:
1. graphical displaying of topological features of each data set in order to define the desired space-time window;
2. synchronization of data sets and standardization of their space-time coordinates.

In details the second type functions are:
- time coordinate standardization: common origin is settled to be that of the least time reference data set;
- space coordinates standardization: a planar thorax mapping symmetrically duplicated with respect to the head position and periodically extended (to maintain contiguity on the right side and on the shoulder) is defined to be the standard space reference. The x - y coordinates are accordingly converted;
- time alignment of data, to eliminate time shifts due to the multiplexer.

The set of data belonging to the chosen window, that we name a parallelepiped, is finally stored as a list formed by four-number elements (v, x, y, t), where x, y, t, are the standardized coordinates and v is the corresponding potential value.

ORDINT - This module can be considered an interface between CPPO and other software systems oriented to display or statistically analyse data disposed on one or two-dimensional regular grid. Its functions are the following:
- to sort the data list (parallelepiped) according to the assigned hierarchical order of x, y, t, variables and the section dimension (one or two). In this way the data are logically divided into subsets, one for each section;
- to compact the list by averaging the possible potential values which have the same x, y, t coordinates;
- to interpolate the possible missing data, in order to get regularly sampled sections. Interpolated values are computed with suitable chosen weighted means.

CONCLUSION

Because of the necessary computing time CPPO is a typical off-line system and therefore its use has to be postponed with respect to the actual acquisition time and the experimental design is previously needed.

It has been constructed for the objectives described in the Introduction and can also be used to define optimal features of other acquisition instruments, oriented to the direct clinical application, and consequently, simpler and less expensive. In fact CPPO makes possible to simulate instrumentations, which are different in probe number and sampling time, and to compare the results, both graphically and statistically.

In this sense the authors are collaborating with a biomedical Italian firm to choose the most convenient features to realize a prototype for monitoring acute myocardial infarction (C.N.R. Special Project on Biomedical Engineering).

REFERENCES

Barone, P., Ciarlini, P., Guspini, A., Macchi, E., Ocello, N.,
        Regoliosi, G., and Taccardi, B., 1978, GISPEM: A Graphic Inter-
        active System for the Processing of Electrocardiographic Maps,
        in: "Modern Electrocardiology", Z. Antalóczy, ed., Akademiai
        Kiadó, Budapest, pp. 127-131.
Cottini, C., Dotti, D., Gatti, E., and Taccardi, B., 1972, A 240-
        probe instrument for mapping cardiac potentials, in:"The elec-
        trical field of the heart", P. Rijlant, ed., Press. Acad.
        Europ., Bruxelles, 99-102.
Ocello, N., 1979, PLOFIL: un sistema grafico per la rappresentazione
        su video Tektronix 4014 di curve piane assegnate per punti,
        Quaderno IAC, serie III, N. 111.

TRIDIMENSIONAL DISTRIBUTION OF HEART POTENTIALS

AFTER ENDOCARDIAL STIMULATION

S. Baruffi, S. Spaggiari, E. Macchi,
G. Arisi, D. Stilli, E. Musso,
R. Th. van Dam, B. Taccardi

Institute of General Physiology
University of Parma
Parma, Italy

The current interpretation of clinical electrocardiograms is based on the solid angle theory. The excitation wavefront is considered to act as a uniform layer of current dipoles which are oriented normally to the wavefront, with the positive poles pointing towards the resting tissue. This theory has been recently challenged by several authors (Corbin and Scher, 1977; Spach et al., 1979; Baruffi et al., 1978), who pointed out, that a new model of intracardiac generator should be developed, where the fiber direction (Corbin and Scher, 1977) and the conductivity along different axes (Spach et al., 1978) should be taken into account. However, it has not yet been established how far these factors affect the potential distribution in the extracardiac conducting media.

With a view to attempting a quantitative approach to this problem, isolated dog hearts were perfused with the blood from donor dogs and placed in the middle of a cylindrical tank filled with Ringer solution at 37°C. An intramural needle with 10 lead points one millimeter apart from one another was inserted into the left ventricular wall. It was used for stimulating the heart at the subendocardial level and for recording the intramural potentials in the wall. Electrocardiograms were also recorded from 600 points regularly distributed in the conducting medium.

No increasing positivity was ever recorded from the lead points along the needle or from those parts of the conducting medium that were located in front of the stimulating electrode while the excitation wavefront was spreading through the wall toward the epicardium. Two maxima initially arose on the left and right sides of the

103

negative area. Thereafter the potential pattern exhibited a central
negativity surrounded by a more or less complete circular rim of positivity. This configuration could be observed not only in the proxi-
mity of the heart but also at a distance of several centimeters from
the epicardium.

Our experiments show that a wavefront spreading from endocardium
to epicardium in the left ventricular wall elicits a potential distribution in the conducting medium surrounding the heart, which differs from the configuration one would expect from the solid angle
theory. The potential patterns we observed are probably due to the
dipoles oriented along the fiber direction having a higher strength
as compared to those oriented normally to the fibers (Corbin and
Scher, 1977). According to other authors, the anisotropy and inhomogeneity of the myocardium and surrounding media may play a major role
in determining the features of the potential field. In any event,
Corbin and Scher's observations, which are in agreement with our
data, should be taken into account when interpreting clinical electrocardiograms, since the usual interpretation, which is based on the
solid angle theory, does not seem to be applicable to our experimental conditions.

REFERENCES

Baruffi, S., Spaggiari, S., Stilli, D., Musso, E., and Taccardi, B.,
    1978, The importance of fiber orientation in determining the
    features of cardiac electric field, in: "Modern Electro-
    cardiology," Z. Antalóczy, ed., Excerpta Medica, Amsterdam,
    pp. 89-92.
Corbin, L. V., and Scher, A. M., 1977, The canine heart as an
    electrocardiographic generator: Dependence on cardiac cell
    orientation, Circulation Research, 41:58-67.
Spach, M. S., Barr, R. C., Miller-Jones, E., Miller III, W. T., and
    Warren, R. B., 1979, Origin of extracellular potentials of
    excitation waves in two-dimensional anisotropic cardiac
    muscle, in: Progress in Electrocardiology, P. W. Macfarlane,
    ed., Pitman Medical, Tunbridge Wells, pp. 105-109.

QUANTITATIVE EVALUATION OF BODY SURFACE MAPS IN

NORMAL AND PATHOLOGICAL CONDITIONS

Z. Antalóczy, I. Préda, Gy. Kozmann
and Zs. Cserjés

2nd Medical Clinic of the Postgraduate
Medical School, Budapest
and Central Research Institute for Physics
Budapest, Hungary

INTRODUCTION

Routine ECG methods provide information on the three main types of changes in the electrical activity of the heart:

a) Changes in electrically active structures of the heart and/ or functional and metabolic disturbances of the electrical activity of the heart.

b) Disturbances in the formation and conduction of impulses.

c) Positional changes of the electrical axis of the heart.

In the present study the multipolar expansion introduced by Geselowitz (1960) was used as a feature extraction for evaluating the information content in normal and in some characteristic pathological conditions. The experimental data were obtained from a relatively small number of patients forming different but homogeneous groups from the clinical point of view and the dipolar and the multipolar content of the body surface maps were determined during the heart cycle.

THEORETICAL BACKGROUND

It is well known that each heart cycle is triggered by the electrical activity which spreads through the ventricular musculature. The time - varying body surface potential distribution is closely

related to this process of activation and recovery. It would be of
great use for clinical practice if the inverse problem could be
solved, i.e. if the determination of the time varying sources from
the surface potential distribution could become possible.

In theoretical studies it has been pointed out that no straight
forward solution of the inverse problem exists (Geselowitz, 1976),
which means that it is impossible to determine a unique source distribution which will account for the measured potential field.

Geselowitz (1960) pointed out that instead of the determination
of the source distribution useful equivalent generators can be calculated from the surface potentials which are related to the sources.
The method of Geselowitz, referred to as the multipolar expansion,
represents the cardiac sources by an infinite series of equivalent
generators. Its first two components are called dipole and quadrupole. The parameters of these equivalent sources can be calculated
by the following equations:

$$a_{11} = X = g \int V dS_x$$

$$b_{11} = Y = g \int V dS_y$$

$$a_{10} = Z = g \int V dS_z$$

$$a_{20} = g \int V(2z dS_z - x dS_x - y dS_y)$$

$$a_{21} = g \int V(z dS_x + x dS_z)$$

$$b_{21} = g \int V(z dS_y + y dS_z)$$

$$a_{22} = \tfrac{1}{2} g \int V(x dS_x - y dS_y)$$

$$b_{22} = \tfrac{1}{2} g \int V(y dS_x - x dS_y)$$

$$(1)$$

where a_{11}, b_{11}, a_{10}:   dipolar components

a_{20}, a_{21}, b_{21}, a_{22}, b_{22}: quadrupolar components

V: body surface potential distribution

x, y, z: Chartesian coordinates of the dS
                              surface elements

g: conductivity

The physical content of the different parameters can be better understood by the study of Fig. 1. The generated potential field is represented by isopotential lines on the surface of a sphere with a radius of one cm.

It is, of course, known that the values of the dipolar components - in contrast to the higher terms - do not depend on the coordinates of the origin used for the multipolar expansion (Geselowitz, 1965).

In our computations that point was chosen as the origin at the time instant of the maximum of the spatial magnitude curve, which minimized the mean square quadratic error between the actually measured body surface potential distribution and that reconstructed from the dipolar and quadrupolar components.

To demonstrate the clinical application of dipolar and quadrupolar representations, measurements were taken from healthy individuals and subjects with different pathological electrocardiological abnormalities.

METHODS

Body surface potential maps were constructed from unipolar lead measurements performed at 120 points on the thoracic surface. For determining the geometrical coordinates of the measuring points a simple equipment was used, and for the determination of heart geometry X-ray pictures were taken from two directions (Kozmann et al., 1978).

As only the potential distribution on the thoracic surface was available, two kinds of truncation strategy were applied for the calculation of the multipolar components:

In the first version, the real geometry of the chest was used and only the upper and lower parts of the body surface were replaced by horizontal closing planes (torso geometry).

In the second version the real body was replaced by a sphere, the measuring points of which were distributed equidistantly according to the azimuth and elevation angles (sphere geometry).

Even though in both geometries the main features of the dipolar and quadrupolar components were similar, the curves obtained from the sphere geometry calculations were usually the more characteristic ones. This was the reason why the latter was selected for the demonstration of the results.

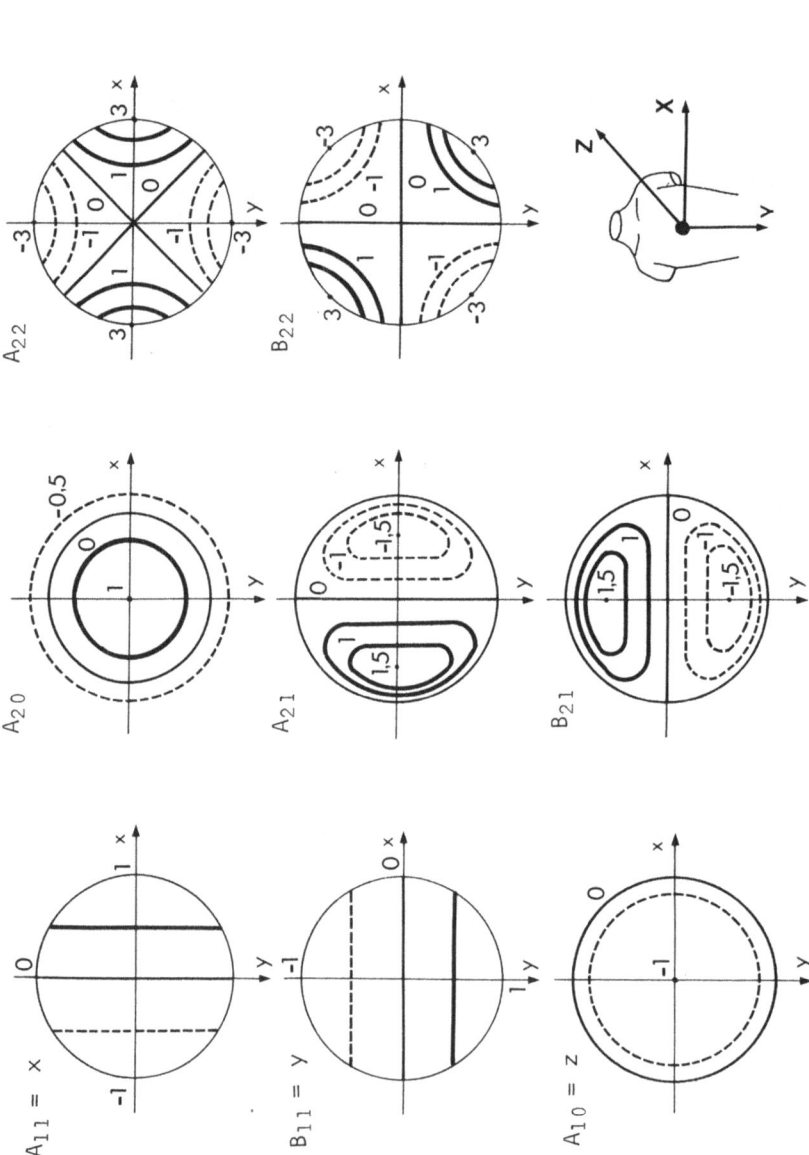

Fig. 1. Potential fields generated by the dipolar (A_{11}, B_{11}, A_{10}) and quadrupolar (A_{20}, A_{21}, B_{21}, A_{22}, B_{22}) components on the surface of a sphere with a radius of one cm.

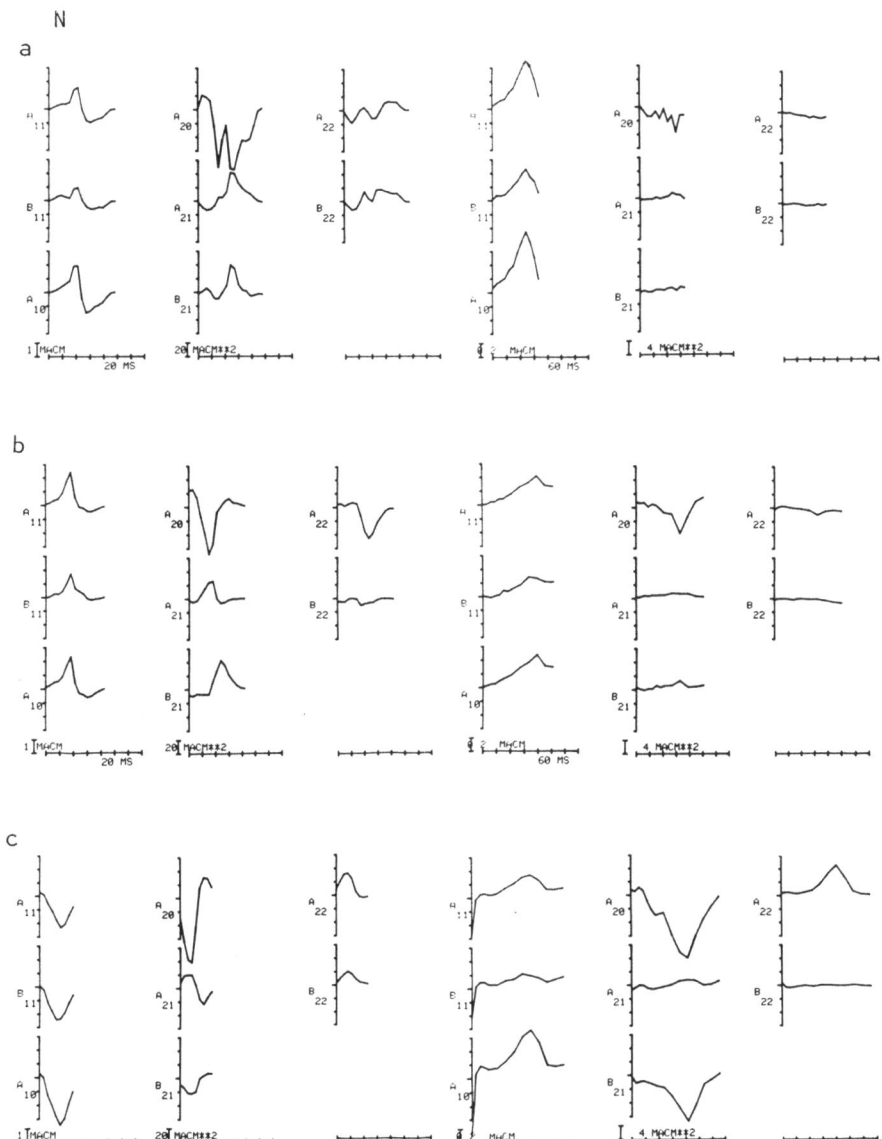

Fig. 2.   Dipolar and quadrupolar components: (a) normal case, (b)
          postero-basal myocardial infarction, (c) postero-diaphrag-
          matic myocardial infarction.

RESULTS

Dipolar and quadrupolar components for a representative healthy person are shown in Fig. 2a. Dipolar components of the QRS complex are in their main features similar to the orthogonal leads of Frank X, Y and Z apart from the missing Q wave in the D_Z, which is probably caused by the application of the truncation method of the lower part of the chest during the computations (Antalóczy et al., 1978). The A_{20}, A_{21} and B_{21} quadrupolar components of the QRS complex have significant values too. In the cardiac repolarization (ST and T wave) the dipolar components are unambiguously prominent, the relative values of the quadrupolar components are very low.

Postero-basal myocardial infarction (PBMI) (Fig. 2b) causes no major differences in the dipolar components of the QRS complex. But the quadrupolar components in the A_{21} and A_{22} subdivisions mainly reflecting the left and upper side of the heart are characteristically different from the normal ones, which show the surface reflection of fall out in the left upper position on the back. The repolarization process causes no significant differences from the normal except for the A_{20} component.

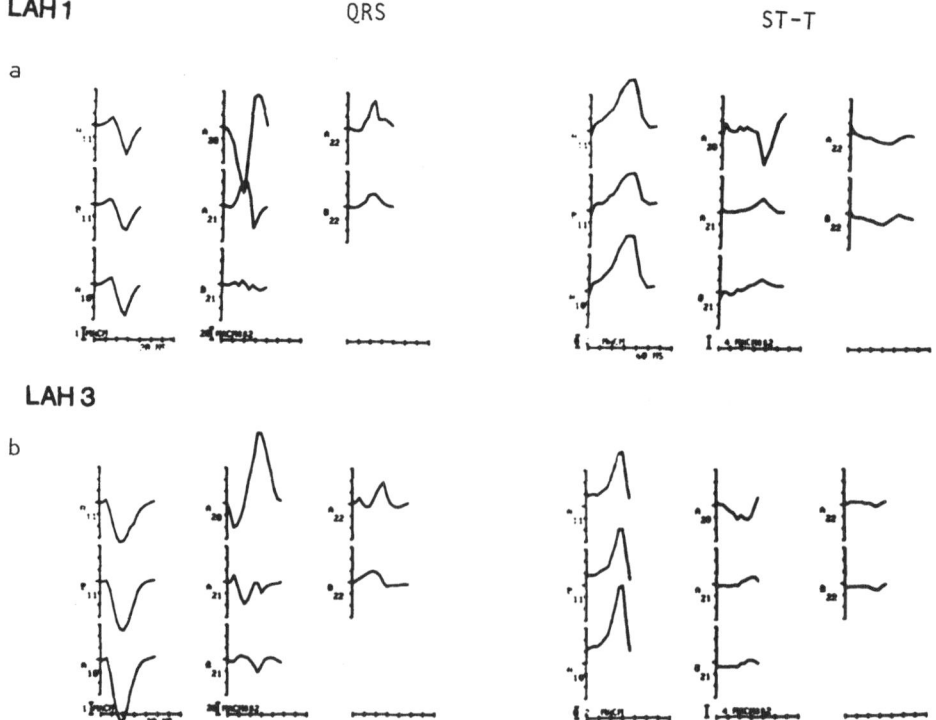

Fig. 3.   Dipolar and quadrupolar components: (a) left anterior hemiblock, (Type 1), (b) left anterior hemiblock (Type 2).

Figure 2c depicts the dipolar and multipolar components of a representative patient suffering from posterodiaphragmatic myocardial necrosis (PDMI). Dipolar components differ characteristically from the normal also as it does for patients suffering from PBMI, both in the time interval of depolarization and repolarization. Quadrupolar components are characteristic for this pathological condition in the B_{21} and A_{22} curves and it is also significant that the amplitudes of these quadrupolar components are visibly augmented during cardiac repolarization.

Two different types of left anterior hemiblocks (LAH) distinguished by the surface mapping method are illustrated in Fig. 3; these are known as Type 1 and 3 (Preda et al., in this volume). With regard to cardiac depolarization, the differentiating features can be explored by the quadrupolar components A_{20} and A_{21}; the positive and negative deflections of A_{21} in the Type 3 LAH can be correlated with the oscillating trajectory of the main maximum of the surface maps towards the left and posterior surface of the chest (Préda et al., in this volume).

In the case of LAH superimposed by posterodiaphragmatic myocardial infarction (PDMI) significant increases are shown in the relative contribution of the quadrupolar components (Fig. 4a); A_{21} reflects the left anterior hemiblock (Type 1), while B_{21} and A_{22} characterize the postero-disphragmatic myocardial necrosis, just as was seen in the case of a similar localization of myocardial infarction without conduction defect (Fig. 2c). The relative values of the quadrupolar components in both depolarization and repolarization in the case of PDMI + LAH are highly augmented in contrast to the extensive posterior myocardial infarction (EPMI + LAH) where the quadrupolar values of repolarization are decreased (Fig. 4b).

Figure 5 depicts the dipolar and quadrupolar components of a pure block of the left bundle branch (Fig. 5a) and the same components when the conduction defect was associated with anterior (Fig. 5b) or chronic PDMI (Fig. 5c). Both localizations of myocardial necrozes cause the dipolar components to diminish during the QRS complex. Distinction of anterior (AMI) and PDMI can probably be based on the form analysis of the A_{20}, A_{21} and A_{22} quadrupolar components which differ from the LBBB cases without myocardial infarction. In contrast to this, during the repolarization the distinction of different forms of myocardial infarction associated with LBBB, may be based on the A_{20}, A_{22} and B_{21} quadrupolar components. The ratio between the quadrupolar and the dipolar components during the QRS complex is higher in both localizations of myocardial infarction if superimposed on LBBB.

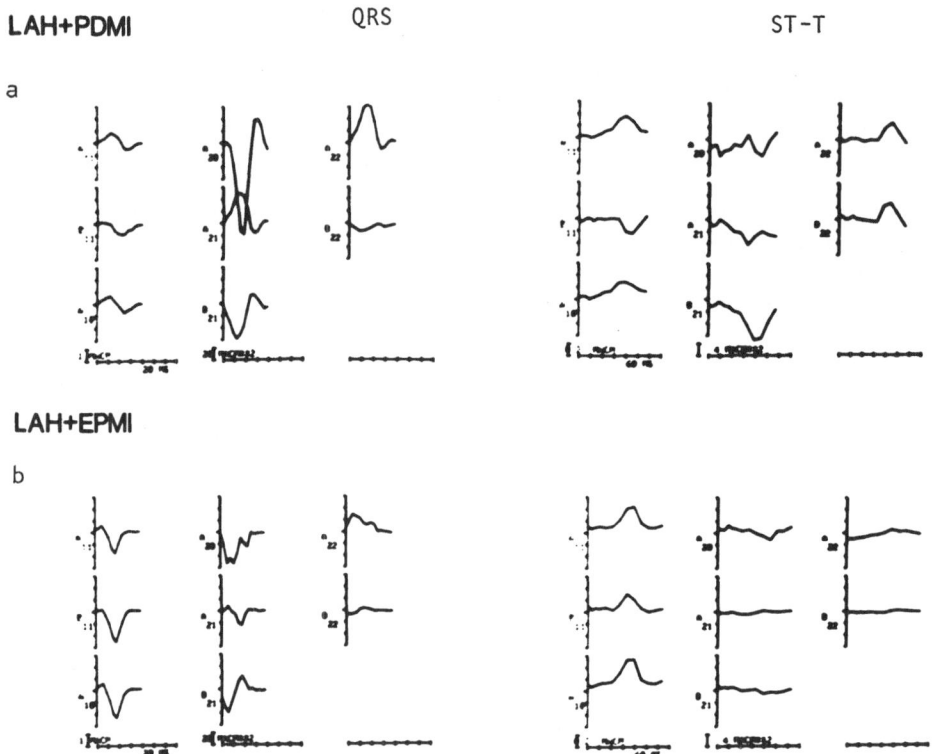

Fig. 4. Dipolar and quadrupolar components: (a) left anterior hemi-
         block superimposed by posterodiaphragmatic myocardial in-
         farction, (b) left anterior hemiblock superimposed by exten-
         sive posterior myocardial infarction.

DISCUSSION

        According to the results of previous investigations of multi-
polar expansion (Antalóczy et al., 1979; Arthur et al., 1972; Guardo
et al., 1976), mostly concerning normal cases, it was demonstrated
that the distribution of potentials over the body surface during almost the whole heart cycle can be expressed with good approximation
by the first member of the multipolar expansion, i.e. by the dipolar
components. This dipolar character can be considered as a result of
partial cancellation of the activating wavefronts of different directions, triggered by the impulse conducting system. This electrical
"balance" or "harmony" which exists originally is characteristic for
the normal cases and can be fundamentally changed by conduction defects and by necrosis of different parts of the left ventricle.

        In earlier reports Antalóczy et al. (1978) have systematically
collected and statistically evaluated the dipolar phenomena in Cartesian and polar representation existing in pathological cases. The

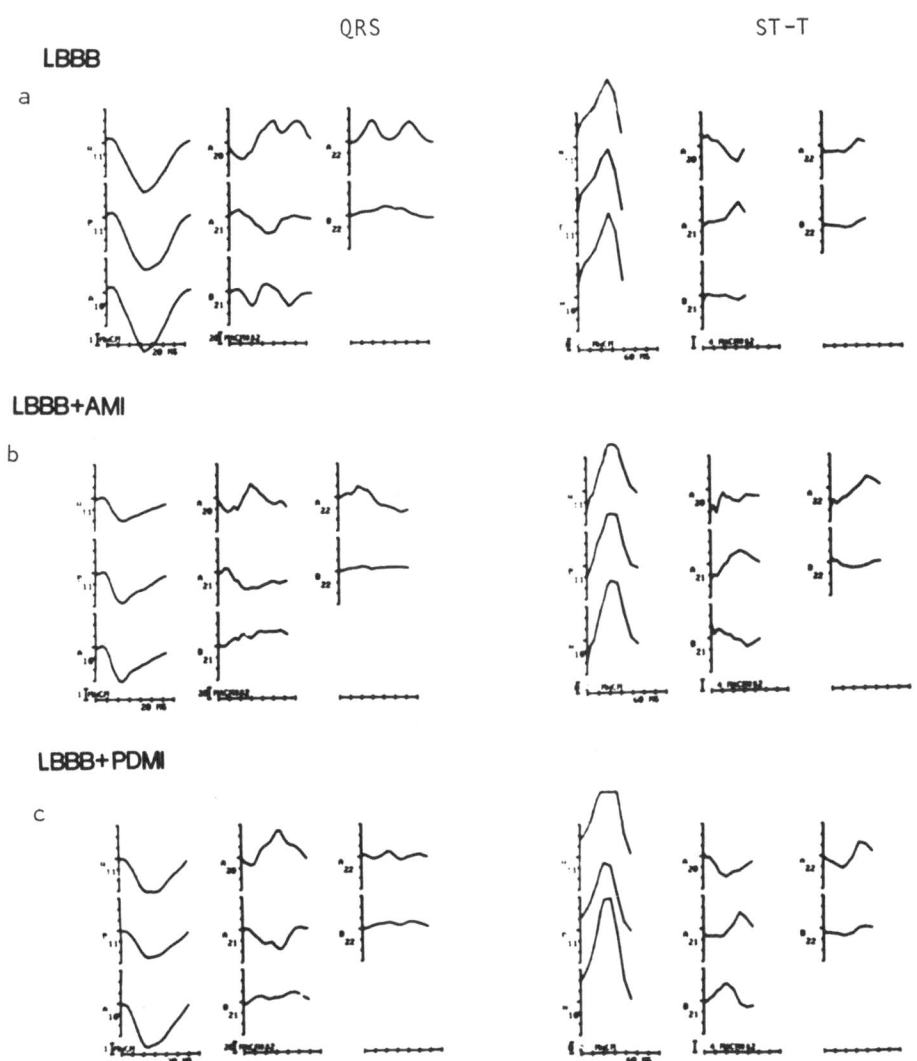

Fig. 5.   Dipolar and quadrupolar components: (a) pure left bundle
branch block, (b) left bundle branch block superimposed by
anterior myocardial infarction, (c) left bundle branch block
associated by chronic postero-diaphragmatic myocardial infarction.

present investigations suggest the clinical diagnostic value of
quadrupolar components to be not only supplementary but to contain
basic information on some pathological cases. By judging our results
this complementary use of quadrupolar expansion might be fruitful,
for instance, in the diagnostics of different localizations of myocardial infarctions superimposed by conduction defects. Further
measurements and their statistical evaluation are needed to confirm
the present results.

REFERENCES

Antalóczy, Z., Strommer, M., Regös, L., 1978, Clinical use of tri-
    axicardiometry, in: "Modern Electrocardiology", Z. Antalóczy,
    ed., Excerpta Medica, Amsterdam, pp. 273-278.
Antalóczy, Z., Préda, I., Shakin, V. V., 1979, Dipolar and quadru-
    polar content of body surface maps evaluated in an infinite
    medium, in: "Progress in Electrocardiology," P. W. Macfarlane,
    ed., Pitman Medical, Tunbridge Wells, pp. 87-91.
Arthur, R. M., Geselowitz, D. B., Briller, S. A., Trost, R. F., 1972,
    Quadrupole components of the human surface electrocardiogram,
    Am. Heart J., 83:663-677.
Geselowitz, D. B., 1960, Multipole representation for an equivalent
    cardiac generator, Proc. IRE, 48:75-79.
Geselowitz, D. B., 1965, Two theorems concerning the quadrupole ap-
    plicable to electrocardiography, IEEE Trans. on Bio-Medical
    Engineering BME, 12:164-168.
Geselowitz, D. B., 1976, Determination of multipole components, in:
    "The theoretical basis of electrocardiology," C. V. Nelson
    and D. B. Geselowitz, eds., Clarendon Press, Oxford, pp. 202-
    212.
Guardo, R., Sayers, McA., B., Monro, D. M., 1976, Evaluation and
    analysis of the cardiac electrical multipole series based on
    a two-dimensional Fourier technique, in: "The theoretical
    basis of electrocardiology," C. V. Nelson and D. B. Geselo-
    witz, eds., Clarendon Press, Oxford, pp. 213-256.
Kozmann, Gy., Préda, I., Shakin, V. V., Szlávik, F., Békési, S.,
    1978, Computer-aided measuring system for complex heart acti-
    vity investigations at body surface and epicardial level, in:
    "Modern Electrocardiology," Z. Antalóczy, ed., Excerpta
    Medica, Amsterdam, pp. 121-126.
Préda, I., Kozmann, Gy., Antalóczy, Z., 1981, Distribution of heart
    potentials on the human thoracic surface in left fascicular
    blocks (in this volume)

A COMPARISON OF ORTHOGONAL LEAD SYSTEM VECTORS USING

DIFFERENT EQUIVALENT DIPOLE GENERATORS

Peter W. Macfarlane

University Department of Medical Cardiology
Royal Infirmary
Glasgow, Scotland

INTRODUCTION

Recently, the hybrid lead system (Macfarlane, 1979) has been introduced and studied (Macfarlane, 1980) using both the Frank (1954) torso model and the lead vector transfer co-efficients of Rush (1975). The hybrid system permits the derivation of both the conventional 12-lead ECG and a corrected 3 orthogonal lead ECG using twelve electrodes from which ten independent leads (I, II, V1-V6, V6R and V neck) are derived. The remaining leads required to form the combined three orthogonal plus 12-lead ECG can be derived mathematically from the ten components using simple equations e.g.

$$avF = II - 1/2\ I$$

$$z = \{1/3\ (V1 + V2 + V3) - V\ neck\} \times .89$$

The Rush model which produces lead transfer co-efficients incorporates fifteen dipole positions within the region of the heart inside the model. One version of the model contains a homogeneous conducting solution while another contains inhomogeneities such as plexiglas rods which permit simulation of torso inhomogeneities. The aim of the present study was to evaluate the hybrid, axial (McFee and Parungao, 1961) and Frank (1956) lead systems with respect to the fifteen different dipole positions.

METHODS AND MATERIALS

The data of Rush provide a set of heart lead transfer co-efficients between each of the fifteen dipoles and 860 electrodes on

115

the body surface. These were obtained for each dipole by orientating the dipole parallel to one of the three axes X, Y and Z and recording the potential at each electrode and repeating for the other two orientations of the dipole. Thus for one dipole position, there are three potential values at each electrode being the heart lead transfer coefficients for that particular dipole and electrode.

In order to calculate the lead vector for a particular lead and dipole the following approach is used. Consider lead X of the hybrid system which can be expressed as follows:

$$X = (V5 - V6R)/C$$

C is a correction co-efficient which makes allowance for the particular model under study. For the homogeneous model it has the value 63.5 and for the inhomogeneous model has the value 39.7 producing results for lead vectors with dimension mV/cm (Rush, 1980). For lead X and dipole 7, it is found that the heart lead transfer co-efficients for V5 are, according to Rush (97, – 16, 5) and for V6R they are (-31, -36, 16). When these data are inserted into the above equation for the inhomogeneous model where C has the value 39.7, the following results are obtained for the X lead vector: (3.22, .50, -.52). In this way the X lead vector for each dipole position can be calculated. Note that the three components refer to the xyz components of the X lead vector and not to the XYZ lead vectors. A similar procedure is adopted for other leads using the appropriate equations for each lead system.

RESULTS

Table 1 shows the results obtained for the values of the lead vectors for all lead systems using selected dipoles for the inhomogeneous model. Table 2 shows the results for the components of the lead vectors for the Frank lead system, using the inhomogeneous model.

DISCUSSION

The striking finding from this study is that the lead vectors have quite widely varying component values and hence strengths. It is of interest to compare the Z lead strengths of the three lead systems for dipole eleven for example which is situated towards the base of the heart. Here it can be seen that the y component of both the Frank and the axial lead system is relatively strong compared to the z component and indeed for the hybrid system the y component is even greater than the z lead component suggesting that all three lead systems are incapable of accurately recording Z lead components from this area of the heart. For the lower region of the heart a typical

Table 1. The components of the Z lead vector for selected dipoles are detailed for the axial and hybrid lead systems. Similar components for the Frank system can be found in Table 2. The inhomogeneous model has been used.

Dipole	Axial			Hybrid		
	x	y	z	x	y	z
1	1.12	0.68	-2.82	1.57	2.19	-6.95
2	0.23	0.41	-2.34	0.13	2.7	-4.12
5	0.05	-0.27	-2.97	0.13	1.12	-3.65
6	-0.57	-0.19	-3.35	2.09	1.8	-4.45
7	0.27	-1.37	-5.49	-0.19	2.05	-10.6
11	0.11	1.78	-2.74	0.30	3.97	-3 .33
13	0.36	0.56	-2.25	0.05	4.32	-2.7
15	0.58	0.72	-4.36	0.88	-0.29	-8.19

Table 2. Components for the XYZ lead vectors for the Frank lead system with respect to the 15 individual dipoles. The inhomogeneous model has been used.

Dipole	Lead X			Lead Y			Lead Z		
	x	y	z	x	y	z	x	y	z
1	2.07	0.11	0.21	0.48	2.05	-0.93	-0.28	-0.44	3.86
2	2.47	0.19	0.54	0.06	2.63	-1.71	0.23	-0.79	2.12
3	2.62	0.44	-0.37	0.79	2.44	-0.77	0.10	-0.29	2.10
4	1.96	0.65	-0.59	1.18	3.43	-1.26	-0.55	-1.06	2.13
5	2.53	0.41	-0.38	0.27	2.71	-0.25	0.12	0.45	2.25
6	3.40	-0.05	-0.99	-0.72	2.09	-0.57	0.54	0.00	3.32
7	3.16	-0.11	-1.22	0.31	1.82	-0.64	-0.61	-0.04	5.07
8	3.43	-2.02	-1.33	-0.93	3.82	0.44	0.20	-0.17	1.38
9	2.23	0.12	-0.31	0.43	3.32	-0.37	-0.06	-0.30	2.18
10	3.51	-0.29	-2.50	-0.09	1.92	-0.36	-1.19	0.47	5.89
11	1.74	-0.76	0.03	0.36	2.07	-0.82	0.27	-1.78	2.17
12	1.87	-0.29	-0.34	0.51	3.67	-1.04	0.03	-1.02	2.00
13	1.30	1.86	-0.13	-0.16	3.89	-0.57	-0.15	-0.45	2.68
14	5.72	-1.64	-4.04	-0.38	2.67	-0.14	-3.88	4.00	7.44
15	3.49	-0.35	0.98	0.24	2.43	-0.16	-0.83	1.14	1.43
Mean	2.76	-0.11	-0.696	0.156	2.73	-0.61	-0.404	-0.02	3.06

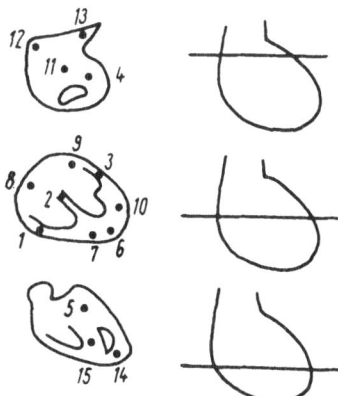

Fig. 1.   The positions of the 15 dipoles are illustrated.   They are situated at three different levels as indicated.

dipole such as dipole 7 (see Fig. 1) has better characteristics with the ratio of z to y components for the Z lead vector being better than 5/1 approximately.   In the apex of the heart e.g. dipole 15, the hybrid system appears significantly better than the axial and Frank lead systems.

The results from this study indicate the complexity of the task of designing a truly orthogonal lead system.   It would seem that the best that can be done is to adopt the compromise approach of Frank where a single dipole in a central region of the heart is used or else perhaps the mean values for the components averaged over the fifteen positions of the Rush model although both approaches are open to criticism.

Use of a corrected orthogonal lead ECG system will always be an approximation at best and inaccuracies in the calculation of the equivalent resultant dipole vector components have to be expected. The design of any lead system is a matter of theoretical attractions versus practical considerations and the hybrid system was evolved from a need to stress the latter in particular.   As shown elsewhere (Macfarlane, 1980), the homogeneous and inhomogeneous transfer co-efficients of the Rush model result in hybrid lead vector strengths which do not compare accurately with axial lead vector strengths derived from the same model when set against actual comparisons under-taken in 276 patients using for example the ratio of R wave ampli-tudes in corresponding leads of different systems.   This has led to the decision that the data of Frank derived from his image surface be retained in the calculation of the relative gains of various leads of the hybrid lead system, as previously published.

ACKNOWLEDGEMENT

The author gratefully acknowledges the most helpful assistance of Professor Stanley Rush of the University of Vermont who provided the correction factors to enable lead strengths to be calculated in mV/cm.

REFERENCES

Frank, E., 1954, The image surface of a homogeneous torso, <u>Am. Heart J.</u>, 47:757-768.

Frank, E., 1956, An accurate clinically practical system for spatial vectorcardiography, <u>Circulation</u>, 13:737-749.

Macfarlane, P. W., 1979, A hybrid lead system for routine electro-cardiography, <u>in</u>: Progress in Electrocardiology, P. W. Mac-farlane, ed., Tunbridge Wells: Pitman Medical, 1-5.

Macfarlane, P. W., 1980, A comparison of axial leads using lead transfer co-efficients, <u>in</u>: Abstracts of 7th International Congress on Electrocardiology, p. 57.

McFee, R., Parungao, A., 1961, An orthogonal lead system for clinical electrocardiography, <u>Am. Heart J.</u>, 62:93-100.

Rush, S., 1975, An atlas of heart lead transfer co-efficients, Vermont: University Press of New England.

Rush, S., 1980, Personal Communication.

PHYSIOLOGICAL INFLUENCES UPON THE DYNAMICS OF SURFACE

MAPS DURING ST-T: EFFECTS OF VARIED HEART RATE

E. Schubert, M. Engst, and K. Mohnike

Institute of Physiology
Humboldt-University
Berlin, GDR

INTRODUCTION

The rapid improvement of electronics and of data processing
facilitates the use of the complex procedure of mapping of the dynamics of the cardiac electric field for clinical purposes. The
interest of physicians in these electrocardiographic mapping investigations increases more and more, because this method belongs to the
noninvasive ones and allows frequently repeated overall field recordings without any serious impairment of the patient. This is of
importance in the clinical investigation of such diseases, in which
a continuous supervision of the cardiac functions of the patient is
needed, for example the recent myocardial infarction (Selwyn et al.,
1978), or an exact localization of disturbances in the cardiac excitation pathways as for example ventricular preexcitations or conduction defects (Préda et al., 1978). It also improves the diagnoses
which utilize follow up studies of the dynamics of the cardiac performance during exercise tests especially in ischaemic heart diseases
(Block et al., 1978; Fox, 1979). Empirical methods for this purpose
have been developed in the form of the Ecg-ST-T-mapping (Fox et al.,
1978). The use of this method generates the necessity to define the
normal pattern of the field maps with its dynamics especially in the
ST-T-period and its variabilities produced by respiration, work load
or other physiological influences.

During the preceding conference on the cardiac electric field
in Smolenice in 1976 we presented some studies on the effects of different inflation of the lungs in the ST-T maps. These studies elucidated a significant but unpresumed strong determination of the
normal field pattern under different respiratory states by the inflation dependent geometrical relationship between the heart and the

chestwall. A minor relevancy is proved for events in the heart, pro-
voked by vegetative influences or other factors (Schubert et al.,
1976).

In order to define other physiological effects on the field map
of the heart further measurements under conditions of varied heart
rate caused by exercise were performed. We expected to demonstrate
with these experiments the regular reactions of the cardiac electric
field generated by the influence of physical load on the heart as a
standard for the estimation of the field map during exercise testing.
The results could possibly be expressed in the terms of changes of
voltage, shapes and temporal shifts of the distribution of potential
maxima, minima and equipotential lines in the field map. Furthermore
with such investigations it will be possible to determinate those
areas of the chest wall, which represent the localization of high and
quickly changing potential gradients thus being extremely sensitive
to the influences of varied heart rate during exercise. These regions should be proposed to be preferential ones for investigating
electrodes even in Ecg-records, as discussed e.g. by Sheffield
(1979).

PROBANDS AND METHODS

The study contains the construction of cardiac electric field
maps of the S-ST-T period from measurements carried out on 16 healthy
male students, 20 - 32 years of age.

100 silver recording electrodes were used as a field matrix
mounted on the chestwall, 80 in 8 rows equidistantially distributed
on the front and 20 in 4 rows on the back. The potentials were derived against the Wilson-CT reference electrode. An additional
Einthoven II lead was used for the time reference. The electrodes
were fixed carefully by rubber strips in the 8 levels between the
shoulder and the lower arch of the ribs. The fixation of the electrodes was ensured by supporting rubber sponges. Electrode paste
served to obtain a tight contact, which was controlled by checking
each Ecg on an oscilloscope screen. Sampling and hold took place by
an operator (Kästner, 1972) consisting of a set of 100 preamplifiers,
one for each electrode with a voltage gain by the factor of 100, a
multiplexer and a main amplifier feeding its output potentials into
a storage oscilloscope. The measuring time for each electrode was
20 µs giving a sampling time of 2 ms for the whole map. After multiplexing and amplification each set of 100 potential values displayed
as a series of digital values on the screen of the storage oscilloscope was photographically recorded and repeated for 2 or 3 beats for
each map. From these registrations the maps were constructed by hand
after optical enlargement using the average values calculated from
the 2 or 3 recordings. 50 µV were the smallest potential steps measured, 20 µV could only be estimated and indicated as positive (+) or
negative (-).

The moments of registration were determined by trigger impulses, derived from the Einthoven reference. The maximum of dU/dt during the upstroke of the R-wave in the reference lead was used as the trigger point. With the help of a delay switch, producing different intervals after the trigger point changeable in steps of 5 ms, the following program of 8 mapping moments was set up:

Map 1: 30 ms after the trigger point at the minimum of S,
map 2: 50 ms after the trigger point at the end of S,
map 3: 60 ms after the trigger point at the beginning of ST,
map 4: 110 ms after the trigger point at the middle of ST,
map 5: 160 ms after the trigger point at the end of ST,
map 6: 220 ms after the trigger point at the upstroke of T,
map 7: 260 ms after the trigger point at the maximum of T,
map 8: 285 ms after the trigger point at the downstroke of T.

This programme was used throughout all measurements. After the exercise test the correspondence to the events in ST and T was altered. The time of map 2 is in conformity with the J-point in the reference Ecg.

All probands were investigated in sitting position on the bicycle ergometer in expiratory state during respiratory arrest. Maps were recorded in the rest state and within 1 min after an exercise test with the intensity of 2 W/kg body weight lasting for 5 min. The exercise increased the heart rate, reaching an average value of 128 beats/min immediately after the test and falling down exponentially to 105 beats/min within 1 min after the end of the exercise. The main heart rates during the measurements after the exercise test were about 110 beats/min. The recordings followed the sequence of the time points from S to T which took nearly 2 min. Therefore after a rest period of 10 min the exercise test was repeated and the maps were then recorded in the reversed sequence from T to S. Thus all registrations used for mapping are situated for sure within the first minute after the end of the exercise.

For ensuring a reliable reproducibility of the measurements the whole procedure of mapping was repeated with 2 probands and single maps were constructed for 3 probands out of 6 to 10 display recordings. The detectable differences in these cases had a maximum of 100 μV.

The results were analysed as features of the series of maps, characterizing the temporo-spatial changes of the field. The statistical evaluation was performed for the absolute values of potentials of the maxima and minima at the moments investigated irrespective to their localization, using the Wilcoxon-test for the discrimination of differences between normal and increased heart rate with an error limit of $p \leq 0.05$.

RESULTS

Similar and reproducible potential patterns during rest and corresponding dynamics under comparably increased heart rate are recorded in all 16 persons. The shape and the localization of the
maxima and minima are generally the same in maps obtained under resting conditions and after the exercise test. Low or nearly no potentials are measured at the electrodes on the back side of the
thorax during the moments of the middle and the end of ST thus pre-
venting the statement of defined potential patterns. No displace-
ments of the areas of positivity or negativity are discovered after
the exercise test as it could be demonstrated before during deep
inspiration (Schubert et al., 1976).

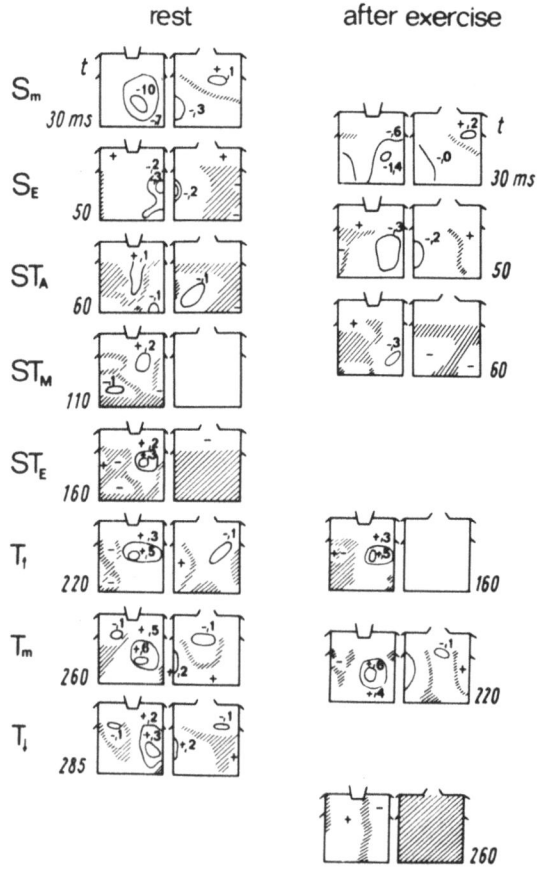

Fig. 1. Cardioelectric field maps of a representative person during
         resting conditions (left column) and during 1 min after the
         exercise test (right column). The indications on the left
         side and the time values correspond to the investigating
         program (see text), the placement of the maps after exercise
         demonstrates the correspondence of the pattern.

The main features of all maps are: (see Fig. 1).

During S there is a large precordial S-negativity, a back S-positivity, and a small positivity beginning in the right axillar which continues up to T.

During ST the potential and area of the S-negativity decreases, and shifts left- and downward and disappears finally. At the same time the T-positivity shifts into the precordial region with a slight change or stability of its potential values.

During T the potential of the precordial T-positivity increases up to T_{max} and remains nearly stationary till it disappears at the end of T, whereas a small second netativity appears in the right axilla. The corresponding T-negativity is situated on the upper part of the back.

Differences between maps under rest conditions and under increased heart rates are found in the duration of the cycle, in the potential values of extrema and in the development of the T-positivity.

The series of events follows faster in the maps after the exercise test. A sufficient temporal correspondence of the characteristic features between maps under resting conditions and after the exercise test can be achieved however by correction of the timing following the QT-shortening caused by the increase of the heart rate (Hegglin et al., 1937). Thus the moment of 260 ms after the trigger point in maps for the resting condition, which characterizes the development of the maximal potential of T, corresponds already with the 220 ms map after the exercise test. In the maps under increased heart rate however, the 260 ms moment does not depict any more detectable potentials at all.

The measurements of the voltages of the extrema show enlarged values in S after the exercise test and in ST during the resting condition. The maximum potential in T shows no significant differences for both conditions.

The follow up of the spatial dynamics of the maximal potential areas throughout the whole S-ST-T-cycle, without respect to their localization, demonstrates changed time patterns in the development of the T-positivity during the ST period (see Fig. 2).

After the exercise the originating T-positivity starts with values of about 100 µV at the end of S which decreases at the beginning of ST to values significantly lower than in the maps during the resting condition. The difference continues up to the moment of the maximum of T. That expresses a prolonged ST-duration with smaller voltages and a steeper upstroke of T after the exercise test caused

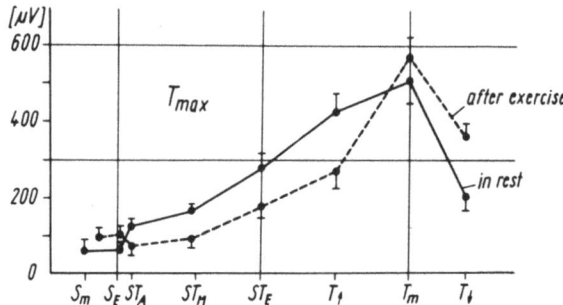

Fig. 2. The potential dynamics of the T-positivity during ST-T
         under resting condition and after the exercise test. Po-
         tential values as $\bar{x} \pm s$ in µV.

by the decrease of the T-positivity at the end of S and its retarded
redevelopment towards the maximum of T. This ST-variation is expressed most clearly by the quotient dV/dt, that means the steepness
of the upstroke of T between the measurements during the resting and
after the exercise test. The relation reaches values of approximately 1.7 for untrained persons out of the group of our probands
(n = 7), increases for students normally active in sports (n = 6)
to 2.7 and reaches a value of 4.1 for the 3 well trained sportsmen of
the group.

     Similar differences exist in the dynamics of the disappearance
of S between the resting condition and the increased heart rate
period (see Fig. 3). The enlarged voltages of the S-negativity
remain for a longer time after the exercise.

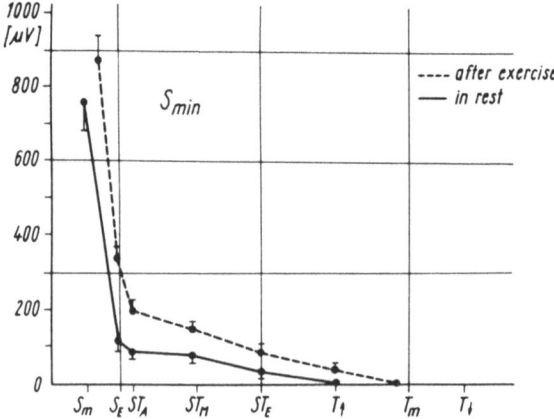

Fig. 3. The potential dynamics of the S-negativity during ST-T
         during the resting condition and after the exercise test.
         Potential values as $\bar{x} \pm s$ in µV.

DISCUSSION

The results of our measurements indicate in general similar recovery patterns of the cardiac electric field during ST-T in the
resting condition and under increased heart rate as it can be detec-
ted by the possible resolution of the voltages. The pattern agreement and the discovered lack of a remarkable ST-negativity is in con-
cordance with results published by Block et al. (1978), who consider
abnormal negativities >50 µV during ST as an indication for pathological processes. The resting ST-T pattern resembles furthermore
the classical ST-T investigations of Amirov (1965), Taccardi (1966
and 1976) and Young (1974). The diverse normal distributions of the
potential during ST reported by Spach et al. (1978) could not be discovered significantly in our small group of probands, especially because of the very small voltages measured during the ST-period.

Inspite of the apparent similarity of the rough time pattern of
the dynamics between the ST-T maps during the resting period and
after the exercise test the changed potential values and the prolongation of ST with the delayed development of the T-positivity demonstrate differences in the recovery of the heart under high heart rate
conditions. Technical variations as mentioned by Simonson (1972) can
be excluded as a source for the differences in our experiments, because all measurements were performed under the same conditions and
repeated exercise tests with 2 of the probands proved a good reproducibility. The "baseline" artifact for high heart rate conditions
as mentioned by Spach et al. (1979) cannot give a sufficient explanation either, the more as the T-positivity starts delayed from lower
potentials during ST to the higher T-maximum (see Fig. 2) with a
delay and a steepness increasing with physical fitness, which means
a decreasing mean heart rate during the exercise test.

The pattern under higher heart rate conditions does not give any
indication of early recovery potentials, which could be demonstrated
easily in many cases during the rest period (Taccardi, 1966;
Schubert, 1972). These missing potentials and the retardation of the
T-development may find an explanation in the well known shortening of
the cellular action potential and the prolongation of its plateau
measured by Trautwein et al. (1962) under increased heart rate conditions. The different degree of the retardation of the T-development in untrained and trained people may possibly refer also to
dromotropic influences, which vary with physical fitness, because of
changing in conduction velocity and the cancellations or summations
of the potentials during the repolarization of the heart. A higher
vagal activity under resting conditions and moderate load is well
known as a consequence of physical training measured e.g. in the
chronotropic regulation of the heart by Eckoldt et al. (1976).

The representation of the difference in the ST voltages and the
T-development is expressed most clearly along the trajectory of the

T-positivity. Taking this and the moments of the interesting events
within the ST-T-period into account, three points can be defined as
reflecting the effects of exercise in the best way (see Fig. 4):

-   the 3rd intercostal space parasternal on the right during the
    end of S for the first decrease of the T-positivity,
-   the midsternal line at the 4th i.c.s. during the middle of ST
    for the smaller voltages during ST, and
-   the 5th i.c.s. medioclavicular on the left at the T-maximum
    for the retardedly developed T.

The trajectory fits well the results reported by Taccardi et al.
(1976) for the movement of the T-positivity reflecting the repolaris-
ation. The region of these three points includes the localizations
of the differential electrodes for the lead cS5 found to be the most
sensitive and specific one for detecting pathognomonic ST-distur-
bances in CAD patients in the exercise-Ecg by Sheffield et al.
(1979). The potential values at the moments defined above in unipolar electrocardiograms from the three localizations found in our
studies may therefore be expected to give the best information about
the "ST-related voltage changes" for the exercise testing. The first
one reflects the known J-point depression, the second one the ST-
depression almost at ST 80 defined as informative for the CAD-diag-
nosis by Block et al. (1978). The third one accentuates the develop-

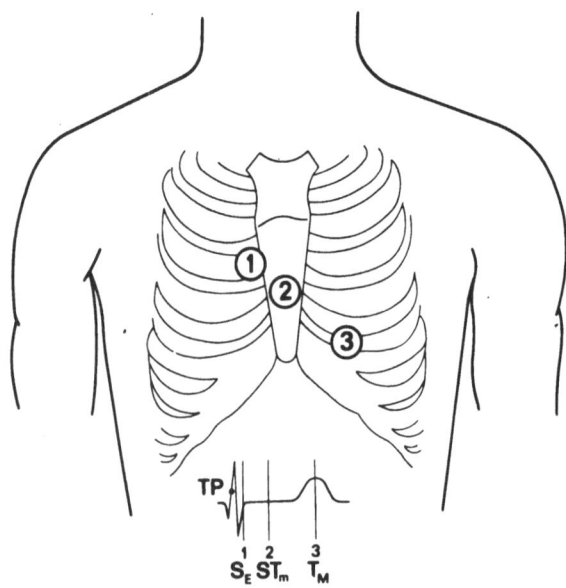

Fig. 4.   Areas and moments of most distinctly expressed ST-T changes
          between the resting condition and after the exercise test.

ment of T which in comparison to the ST-voltage delivers further information about influences of exercise on the repolarization period and should induce studies of these effects for CAD-patients.

REFERENCES

Amirow, R. S., 1965, "Elektrokardiotopographia," Medgis, Moscow, p. 22 f (russ.).

Block, P., Raadschelders, J., Smets, Ph., Darquenne, H., Lanaers, A., Bourgain, R., and Kornreich, F., 1978, Usefulness of Ecg, Vcg and body surface mapping technique during exercise for the diagnosis of coronary artery disease (CAD), in: "Modern Electrocardiology," Z. Antalóczy, ed., Akademiai Kiadó, Budapest, pp. 541-544.

Eckoldt, K., Bodmann, K.-H., Cammann, H., Pfeifer, B., and Schubert, E. (1976), Sinus arrhythmia and heart rate in hypertonic disease, in: "Adv. Cardiol.," H. Abel, ed., Karger Basle, 16:366-369.

Fox, K. M., Selwyn, A. P., and Shillingford, J. P., 1978, A method for praecordial surface mapping of the exercise electrocardiogram, Brit. Heart J., 40:1339-1343.

Fox, K. M., Selwyn, A. P., and Shillingford, J. P., 1979, Projection of electrocardiographic signs in praecordial maps after exercise in patients with ischaemic heart disease, Brit. Heart J., 42:416-421.

Hegglin, R., and Holzmann, M., 1937, Die klinische Bedeutung der verlängerten QT-Distanz (Systolendauer) im Elektrokardiogramm, Z. Klin. Med., 132:1-5.

Préda, J., Bukosza, J., Kozmann, G., Shakin, V. V., Székely, A., and Antalóczy, Z., 1978, Distribution of heart potentials on the human thoracic surface in the cases of left bundle branch blocks, in: "Modern electrocardiology," Z. Antalóczy, ed., Akademiai Kiadó, Budapest, pp.115-120.

Schubert, E., Engst, M., Kästner, R., and Mohnike, K., 1976, Averaged repolarization field maps of healthy persons in different respiratory states, in: "Proceedings of the Conference on Measuring and Modelling of the Cardiac Electric Field," P. Kneppo, ed., Bratislava (in press).

Schubert, E., 1972, The temporo-spatial evolution of activation and repolarization of the heart and its relation to the electric field, in: "The electrical field of the heart," P. W. Rijlant, ed., Press Acad. Europ. Brussels, pp. 506-511.

Selwyn, A. P., Fox, K., Welman, E., and Shillingford, J. P., 1978, Natural history and evaluation of Q waves during acute myocardial infarction, Brit. Heart J., 40:383-387.

Sheffield, L. T., Roitman, D., and Kansal, S., 1979, Computer measurement of bipolar and unipolar exercise Ecg leads, in: "Progress in Electrocardiology," P. W. Macfarlane, ed., Pitman Medical, Tunbridge Wells, pp. 312-315.

Simonson, E., 1972, Physiological variations in the genesis of the
    electrocardiogram, in: "The electrical field of the heart,"
    P. W. Rijlant, ed., Press, Acad. Europ. Brussels, pp. 523-531.
Spach, M. S., and Barr, R. C., 1978, Isopotential mapping in subjects
    of all ages: an analysis of low level potentials, in: "Pro-
    gress in Electrocardiology," P. W. Macfarlane, ed., Pitman
    Medical, Tunbridge Wells, pp. 225-227.
Taccardi, B., 1966, Recent data on the cardiac electric field, in:
    "Neue Erg. Elektrokardiol. I," E. Schubert, ed., Fischer
    Jena, pp. 23-29.
Taccardi, B., de Ambroggi, L., and Viganotti, C., 1976, Body surface
    mapping of heart potentials, in: "The theoretical basis of
    electrocardiology," C. V. Nelson, and D. B. Geselowitz, eds.,
    Clarendon Press, Oxford, pp. 436-466.
Trautwein, W., Kassebaum, P. G., Nelson, R. M., and Hecht, H. H.,
    1962, Electrophysiological study of human heart muscle,
    Circ. res., 10:306-312.
Young, B. D., Macfarlane, P. W., and Lawrie, T. D. V., 1974, Normal
    thoracic surface potentials, Cardiovasc. Res., 8:187-193.

PROBLEMS OF COMPARATIVE ELECTROCARDIOLOGY CREATED BY THE

PROGRESS IN COMPUTER MAPPING OF THE CARDIOELECTRIC FIELD

M. P. Roshchevsky

Institute of Biology
Acad. Sci. of the USSR
Komi Branch
Syktyvkar, USSR

In comparative physiological investigations of the electro-genesis of the myocardium the study of the distribution of the excitation wave in the ventricle walls performed by direct methods is of particular interest. We investigated this process in different classes of animals.

The studies of the myocardial activation and its peculiarities were carried out with a multipolar technique. The results showed that in the most parts of the heart of fishes the excitation is characterized by a synchronous distribution of the depolarization wave with a saw-edged front along the ventricle walls. The excitation begins in the areas adjacent to the atrio-ventricular funnel on the left side of the heart. In the ventricular myocardium the activation wave starts to spread from the zones of depolarization on the left side towards the left lateral and caudo-dorsal walls. In some fishes the zone of depolarization spreads up to the central and right areas of the ventricular basis. From the excited areas the wave of depolarization moves successively to the central areas of the lateral ventricle walls and the apex of the heart. The data obtained are a good evidence for the myogenic character of the spread of the excitation in the ventricular myocardium of the fishes.

The primary regions of excitation in reptiles (turtles and varanus) are situated in the subendocardium of the ventral areas of the ventricular basis close to the atrioventricular orifice. Depolarization of the main mass of the ventricular myocardium takes place as a consequence of the movement of the depolarizing wave in the direction from the primary regions of negativity along the ventricular walls to the apex of the heart and its simultaneous conduction in an acute angle to the epicardium.

Birds show primary regions of excitation in the subendocardium
of the right and left sides of the septum. Together with this an
early negativity is observed in the central subendocardial and sub-
epicardial areas of both of the free ventricular walls. At this
stage a lot of small "spots" of negativity appear inside the myocardium; they spread mosiacally up to the epicardium. From the numerous regions of activation the depolarization wave begins to spread
radially in all directions. In a short time the main mass of the
tissue of the free ventricular walls and the lower thirds of the
septum become excited. The excitation of the lower two thirds of the
septum is caused by the activation waves, which are countercurrent to
each other and spread out from both endocardial surfaces. The upper
third (basis) of the septum is the last to be activated. Its excitation spreads mostly from the basis to the apex. This pattern of the
excitation in the ventricles of the birds can be explained by the
distribution of the Purkinje fibres in the myocardium.

The dog shows primary regions of excitation in the zones of subendocardium (sometimes as deep as 4 mm) at the border of the lower
and middle third of the left side of the ventricular septum. Simultaneously or with a delay of 5 milliseconds two subendocardial zones
on the right side of the septum are excited: one under the basis of
the anterior papillary muscle and one close to the right anterior
free ventricular wall on the level of the anterior papillary muscle.
After this about one third of the diameter of the free ventricular
walls is excited by one flash of depolarization. After that a successive movement of the excitation wave to the epicardium is observed. Such a type of activation in dogs is well explained by the
distribution of the Purkinje-conducting-system in the myocardium of
the ventricles and the architecture of the heart.

In reindeers the primary regions of excitation are situated in
the subendocardium in caudal and cranial areas of the middle third
of the left and in cranial parts of the apical third of the right
side of the ventricular septum. During 5 - 10 milliseconds the subendocardium of the ventricles becomes totally excited by the depolarization wave. The period from 10 - 20 milliseconds is characterized
by the spreading of the depolarization wave from multiple spots of
excitation into all directions. The last to be excited are the
apical zones of the exit cone of the pulmonary artery (40 - 50 milliseconds) and areas of the ventricular septum adjacent to the basis
of the ventricles (45 - 50 milliseconds).

So the data given in this report show that the wave of depolarization moves in fishes along the walls of the ventricle, which might
be due to the absence of conducting system elements in fish myocardium. Reptiles as well as tailless amphibia preserve a successive
type of depolarization of the myocardium. Raptorials and carnivores
are characterized also by a successive type of activation but in
distinction from reptiles they show an almost simultaneous depolari-

zation of subendocardial layers composing the third part of the diameter of the free walls. As a result of this the period of blood
output from the ventricle cavities is shortened. The ungulates are
characterized by a "flashed" type of the ventricular myocardial activation, which results from the excitation of the main mass of the
ventricular myocardium, occurring during a short interval of time.
Considering the chronotopographic distribution of the excitation from
an evolutionary point of view vertebrates show different types of
ventricle activation.

The results of our histological investigations allow us to come
to the conclusion that the presence of conducting fibres of different
diameters both in subendocardial and intramural layers of the myocardium and also close to the epicardium forms a structural basis
for the simultaneous appearance of multiple points of negativity
during the phase of fast depolarization in ventricles of birds and
ungulates.

We also studied the process of intramural excitation in the
atria. The investigated animals (dogs and sheep) did not show any
strong variations in the process of the excitation in both atria.
The excitation wave with a saw-edged front moves from the sinus node
along the walls of the atria. At first, due to the radial distribution of the excitation wave, the areas located close to the sinus
node are depolarized. Then with the radial distribution of the
spreading of the excitation wave we observe a preferable spreading
of activation along the internodal branches of Wenckebach, Thorel,
and Bachmann. The right atrium is excited earlier than the left.

The success achieved by comparative electrocardiology completes
a period of investigations and the data obtained open new horizons in
the study of bioelectrical phenomena in the heart and on the body
surface.

Today it is clear that the investigations of the cardioelectric
field cannot be limited only to studies of depolarization by isochronic maps of the intrinsic deflection or local bipolar potentials
or by equipotential maps of surface activity easily achieved with the
aid of a computer. We can say that in the first case modern techniques will give the possibility to an easy connection of the successive and at the same time separated elements of the ECG in different leads with definite processes of myocardial depolarization. In
the second case comparative electrocardiology is not only faced with
the problem of a possible application of a general field theory to
the description of the processes observed and their modelling for the
myocardium but also with the problem of experimental tests of these
hypotheses with simultaneous registrations of intramural equipotential maps.

There are some possibilities for investigating the process of repolarization in the myocardium too. To look for possibilities to solve this problem seems to be of great importance.

The results of comparative electrocardiology done for the purpose of modelling of the bioelectrical processes show interesting variations for different heart types: for example for hearts with various numbers of ventricles, or hearts with various numbers of conducting elements. It is possible to neglect not only the influence of the body shape but also, quite like as in the isolated heart, the influence of the body itself on the form of the cardioelectric field; for instance in the varan whose heart is practically situated directly beneath the skin.

The mapping of the cardioelectric field puts forward the problem of the synchronization of isochronic depolarization maps and iso-potential instantaneous maps with the contractile function of the myocardium and with certain biochemical phases of ion activity in the myocardium cells. It becomes possible to check experimentally almost any mathematical or biophysical model for its universality, because objects of investigation exist with different types of myocardial ventricular activation. This is the contribution the comparative electrocardiological approach has for the solution of direct and inverse problems in electrocardiology.

THE ELECTRICAL RESISTIVITY OF LUNG TISSUE FILLED

EITHER WITH AIR OR WITH FLUID

Th. Eifrig and H. Schwartze

Department of Pathophysiology
Medical School
Karl-Marx-University
Leipzig, GDR

INTRODUCTION

The cardiac electrical field at the body surface, produced by the heart generator, is decisively influenced among others by the electrical resistivities of the thoracic tissues. Probably, the lungs play a special role by surrounding and shielding the heart, their resistivity being 4 times that of the myocardium (Rush and Nelson, 1976). Measurements of the resistivity of lung tissue in adult dogs resulted in different values depending on the degree of aeration of the lungs (maximal inspiration versus maximal expiration (Rush et al., 1963)). This result could be seen in conformance with certain changes in the cardiac electrical field of human adults, which appear as a function of different respiratory phases. Concerning the lungs, the underlying mechanism is presumed mainly to be found in hemodynamic effects, resulting in an increase of electrical conductivity by an increase of pulmonary blood flow during deep inspiration, the amount of air within the lung tissue supposed to be of less importance (Ruttkay-Nedecký, 1976). The prenatal development of mammals seems to provide a natural model for the study of the mechanisms underlying the physiological respiratory variability of the cardiac electrical field. Among the specific peculiarities of the foetal period, the fluid filled, not aerated lungs give an experimental material, the properties of which make it quite different from lungs of normally breathing subjects. Foetal lungs contain small amounts of blood because about 10% of the total blood volume only reaches the lungs whereas 90% is bypassed through the ductus arteriosus (Dawes, 1968). However, foetal lung alveoli are filled with lung liquid, a homogenous fluid, the volume of which is approximately the same as the functional residual capacity after breathing is estab-

lished (Normand, 1968). It might be expected that measurements of
the electrical resistivity of foetal lung tissue would lead to different results than those of ventilated lungs cleared from the foetal
lung liquid, but containing a greater volume of blood, and an additional volume of air.

MATERIALS AND METHODS

We used guinea pigs, 45 juvenile animals, ranging in age from 3
to 21 days and 30 foetuses of the last quarter of gestation. Experiments have been successfully performed on 36 normally aerated guinea
pig lungs and 26 foetal lungs filled with lung liquid. The lungs
were prepared out of the animals after ligation of the trachea, so
that no air could escape after opening of the thoracic cage. In the
case of the foetuses, only the following procedures had to be performed: the hilum of the left lung had to be tied up and cut and the
surface of the lung tissue carefully cleaned from the blood by a
piece of gelatin spoon. The uninjured lung was brought into a warm,
moist chamber and placed on a little operating table with its outer
surface turned to the measuring electrode. Measurements were performed with a four-electrode technique, similar to that used by Rush
and co-workers (1963). Our setup consisted of 4 electrodes which
were placed on the surface of the tissue. A control-current (I) was
applied on and then removed from the tissue and the resulting curve
being measured by means of two "current" electrodes, and the resulting potential difference (V) was measured between two points on the
tissue surface with two additional "potential" electrodes (Fig. 1).
Under the condition that

   i) the tissue is a homogenous isotropic conductor with a plane
     surface and

   ii) the extent of the sample is large, compared to the electrode
     spacing

the following relation is valid:

$$V = \rho \cdot \frac{I}{2\pi d}$$

This equation is strictly valid only for samples of infinite extent.
However, if the distances of the boundaries are 5 times larger than
the electrode spacing, the corrections are already less than 1%. The
technical arrangement consists of 4 platinum electrodes of 0.1 mm in
diameter and mounted 1 mm distance one from another on a straight
line in a plastic box. From a constant current source d.c. impulses
with I = 25 µA and a duration of 10 and 15 ms were conducted through
the tissue. The chosen impulse duration should simulate a QRS like
current load of the tissue. By means of a two beam oscilloscope the

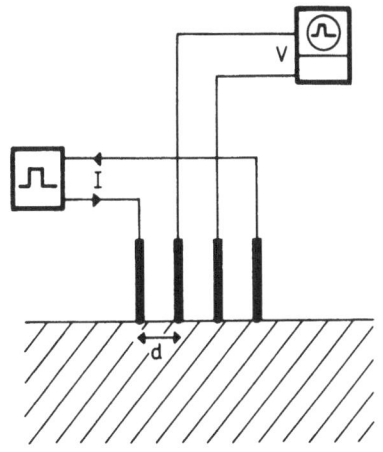

four-electrode technique

assumptions:

i) homogenous, isotropic
   conductor with plane
   surface

ii) sample of large extent
    (in comparison with electrode
      spacing)

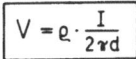

$$V = \varrho \cdot \frac{I}{2\pi d}$$

Fig. 1.   Schematic representation of the measuring circuit

corresponding voltage impulse was recorded and the current impulse
was checked additionally. The accuracy of the measuring equipment
was tested by measurement of a standard solution. The error proved
to be less than 3%.

RESULTS

   i) We compared the shapes of the voltage and current impulses in
order to estimate the magnitude of the capacitive portion of the
specific tissue resistance. We found that in the case of impulses of
10 and 15 ms duration (corresponding to the foetal and the infant
QRS-complex of guinea pigs, respectively) the tissue resistance is
predominantly determined by its ohmic portion, so that the capacitive
portion can be neglected.

   ii) The measured resistivity values are in the same order as
those determined by Rush et al. (1963) for lung tissue of dogs.

   iii) Different physiological states of the lungs cause different
values of tissue resistivity. ϱ equals 1696 ± 235 Ωcm in foetal un-
ventilated lungs, and 1910 ± 239 Ωcm in aerated infant lungs. The
difference proved to be statistically significant at the 1% level.

DISCUSSION

   As consequence of our tissue preparation, the infant lungs contained the ventilation volumes at the end of the expiration. Their
specific resistances, measured on the outer surface far from the
hilum region, nearly equals the resistivity value of adult dog lungs

Table 1. Tissue resistivity of lungs in
different physiological states

guinea pig	$\varrho\,[\Omega cm]$	σ	adult dog (Rush, Abildskov McFee; 1963)	$\varrho\,[\Omega cm]$	σ
fetal lung filled with liquid	1696	12%	—	—	
infant lung aerated	1910	12%	lung maximal expiration	1950	17%
			lung maximal inspiration	2390	17%

in deep expiration (table 1) thus proving that the specific resistance is independent of the two different species (dog or guinea
pig) and the age of the lung donor (adult or juvenile). The greatest
difference between our results and the data of other authors (Rush
et al., 1963) is found between foetal lung tissue, completely without
air, and dog lungs containing the maximal air-volume during deep inspiration, the last of which had a much higher resistivity (see table
1).

The first idea we had regarding this comparison is that the
degree of the ohmic resistance should depend on the ventilation
volume within the lungs. It seems to increase with the amount of
air, which is electrically almost totally insulating. We must, however, consider that changes of the electrical conductivity of lung
tissue are principally determined by two variables acting reversely:
insulation by the air and improvement of the conductivity by the
fluids, e.g. foetal liquid or blood. We may therefore expect a very
high lung conductivity in the foetus which really could be measured.
In ventilated lungs, the combined effects of air volume and blood
flow can be measured only, since under physiological conditions of
the blood-circulation in adults, variations of the two functions,
respiration and blood perfusion, are strictly related to each other.
Changes of the tissue resistivity attributed to changes of the air
volumes in the lungs, which may precisely be produced in experiments, cannot be delimitted from simultaneously increasing blood
volumes during deeper inspiration and vice versa. Experimental division of the two effects might be possible under one particular
physiological condition: when lung circulation develops gradually
depending on age during the postnatal transitional period. The
physiological course of the transitional circulation period in the
neonatal lung is characterised by three processes of different
duration: the very rapid complete aeration within a few minutes
after the onset of respiration; the promptly beginning clearance of

the lung from the liquid, which is completed after a few hours; and the delayed readjustment of the lung blood flow which nearly takes the full time of the transitional circulation period (Dawes, 1968).

In conclusion, we would like to argue two possibilities of further investigation concerning the problem of the electrical lung resistivity:

i) Measurements of the resistivity of the lung tissue of guinea pigs of a defined postnatal age within the transitional circulation period, the lungs containing a defined air volume, in order to estimate the influence of blood filling.

ii) As a control, measurements of the resistivity of foetal lung tissue artificially ventilated in order to estimate the influence of air filling without altering the fluid volume.

REFERENCES

Dawes, G. S., 1968, Foetal and Neonatal Physiology, "Year Book," Medical Publishers, Chicago.
Everett, N. B., and Simmons, B. S., 1954, The magnitude of increase in the pulmonary blood volume of the postnatal guinea pig, Anat. Rec., 119:429-434.
Normand, I. C. S., 1968, The uptake of liquid from the lungs of the foetus, Proc. Roy. Soc. Med., 61:290-291.
Rush, S., Abildskov, J. A., and McFee, R., 1963, Resistivity of body tissues at low frequencies, Circulation Res., 12:40-50.
Rush, S., and Nelson, C. V., 1976, The effects of electrical inhomogeneity and anisotropy of thoracic tissues on the field of the heart, in: "The Theoretical Basis of Electrocardiology," Clarendon Press, Oxford, pp. 323-354.
Ruttkay-Nedecký, I., 1976, Effects of respiration and heart position on the cardiac electric field, in: "The Theoretical Basis of Electrocardiology," Clarendon Press, Oxford, pp. 120-134.

ESTIMATION OF CARDIAC EXCITATION ON THE BASIS OF STIMULUS

RESPONSE FUNCTIONS AND EPICARDIAL ACTIVATION ISOCHRONES

Gy. Kozmann and I. Préda

Central Research Institute for Physics
and Postgraduate Medical School
Budapest, Hungary

At a given instant the potential distribution on the body surface is produced by the cells just being depolarized. Even presuming
the full knowledge of the body surface potential field it is difficult to retrace the activation process, because this is characterized
by the resultant of wave-fronts propagating from different focuses.
The interpretation is usually easier if the distribution of the potential and/or current fields on the epicardium are known. The
fields generated by the separated wavefronts are superimposed here
as well, but their contributions are weighted differently. Though
the knowledge of the epicardial fields enables us to obtain exact
information on the epicardial activation, the excitation of the
deeper layers - especially that of the septum - remains more or less
unknown (Taccardi et al., 1972).

In the present paper:

i) A field decomposition method is outlined which is based on
the knowledge of the stimulus - response functions (SRF)
corresponding to the excitation of the main branches or
fascicles of the impulse conduction system. An estimation
of the most important SRFs is given.

ii) With the help of a simplified anisotropic myocardial model
the influence of some anatomic and conduction velocity
parameters on the epicardial and intramural activation is
analysed.

iii) The results obtained are used for the interpretation of
experimental data derived from dogs at normal temperature
and during hypothermia.

ESTIMATION AND USE OF BODY SURFACE STIMULUS-RESPONSE FUNCTIONS

Using the terminology of the system theory the ϕ_N body surface
potential field is nothing else than a stimulus (or impulse) response
function of a one-input multiple-output system, where from a physical
point of view the His bundle is the input, and the measuring points
on the chest surface are the output terminals.

It is known that the His bundle is continued in the right and
left bundle branches which divide further usually into three main
fascicles, etc. This being the case we may speak of further stimulus-response functions, too. Namely, if the system is considered as
a two-input or a six-input system the ϕ_R right and ϕ_L left bundle
branch responses or the different right anterior (ϕ_{RA}), septal (ϕ_{RS}),
diaphragmatic (ϕ_{RD}) and left anterior (ϕ_{LA}), septal (ϕ_{LS}), posterior
(ϕ_{LP}) fascicular response functions can be defined. At least in the
period of the early ventricular depolarization we may suppose that
the principle of superposition can be applied to the response functions defined above,

$$\phi_N(t) = \phi_L(t-t_L) + \phi_R(t-t_R) =$$

$$= \left[\phi_{LA}(t-t_{LA}) + \phi_{LS}(t-t_{LS}) + \phi_{LP}(t-t_{LP}) \right] \qquad (1)$$

$$= \left[\phi_{RA}(t-t_{RA}) + \phi_{RS}(t-t_{RS}) + \phi_{RD}(t-t_{RD}) \right]$$

where: t_L and t_R: conduction time along the His bundle

$t_{LA} \cdots t_{RD}$: stimulus conduction times from the His bundle to
the LA,...RD main fascicles.

Later on, when the originally separated activation wavefronts
merge a more complicated equation will be true.

We are convinced that not only from the theoretical but also
from the clinical point of view it would be useful to have a decomposition of the body surface field in terms of the different SRFs
defined previously.

For experimental determination of the bundle branch or fascicular level SRFs needed for the decomposition, we made use of the
following working hypothesis: when averaging the corresponding body
surface potential maps of patients with normal or pathological conduction types, the specific characteristics will disappear (diminish), the common, basic features remain. The specific characteristics are due to the actual realization of the conduction system,
to the deviations in the conductive and geometrical properties of
the volume conductor surrounding the heart, etc. If our hypothesis
is true then the SRFs can be calculated from the averaged maps of

homogeneous normal (N), left anterior hemiblock (LAH), left posterior hemiblock (LPH) and left septal fascicular block (LSB) patients according to such simple equations as:

$$M\left[\phi_{LA}\right] = M\left[\phi_N\right] - M\left[\phi_{LAH}\right]$$

$$M\left[\phi_{LS}\right] = M\left[\phi_N\right] - M\left[\phi_{LSB}\right] \tag{2}$$

$$M\left[\phi_{LP}\right] = M\left[\phi_N\right] - M\left[\phi_{LPH}\right]$$

where: $M[\ ]$: the mean value of the matrix time-functions in brackets.

ϕ_{LA}, ϕ_{LS}, ϕ_{LP}: left anterior, left septal and left posterior fascicle SRFs.

Our results are demonstrated in Fig. 1, where the time course of left fascicular SRFs is represented.

A qualitative proof of the results obtained is that: the characteristics of the maps expected on the basis of the known anatomic location of the regions stimulated by the main fascicles correlate well with those of the measured maps.

In consequence of the averaging method used, it is implicit that a certain sequence of the stimulation of the different endocardial regions is presumed. During decomposition however, it would not be wise to apply always such a rigid restriction. Making use of our results in the form of Eq. 3, and by reasonably modifying τ_i and τ_j values, such individual variations as the deviations from the mean conduction velocity or changes in the heart geometry, etc. can be taken into account.

$$\phi(t) = \sum_{(RV)} k_i \phi_i(t-\tau_i) + \sum_{(LV)} k_j \phi_j(t-\tau_j) \tag{3}$$

where $k_i \begin{cases} = 0, \text{ if the } i^{th} \text{ right ventricular fascicle is blocked} \\ = 1 \text{ otherwise.} \end{cases}$

$\quad\quad k_j \begin{cases} = 0, \text{ if the } j^{th} \text{ left ventricular fascicle is blocked} \\ = 1 \text{ otherwise.} \end{cases}$

A SIMPLE ANISOTROPIC MODEL TO ESTIMATE HEART MUSCLE ACTIVATION

Until now the mechanism of activation was totally disregarded, only the input-output relationships were investigated. In this section we try to better understand the activation propagation in the myocardium by using simplified models.

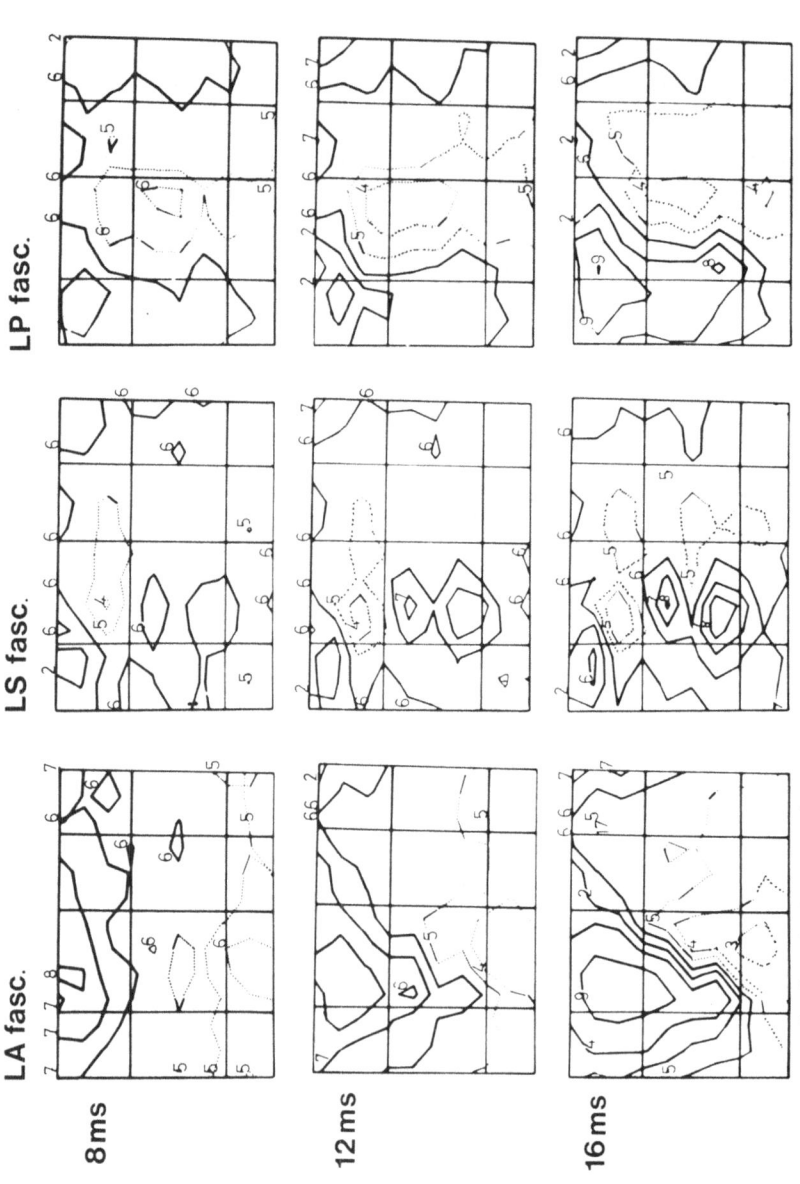

Fig. 1. Time-course of the left fascicular stimulus-response functions. The time instants indicated are measured from the Q wave onset. The left side of the maps represents the anterior, the right side the posterior chest surface. The isopotential lines of the positive potentials are drawn by continuous and that of the negative potentials by dotted lines. Line No. 6 stands for zero potential. The other isopotential lines were drawn at the amplitude values of ±100; ±200; ±400 and ±600 μVs.

Through experimental evidence it became obvious that even in a simplified treatment the anisotropy of excitation propagation has to be taken into account (Baruffi et al., 1977; van Oosterom, 1978). Furthermore, as the anisotropy is closely related to the fiber orientation the main properties of the heart muscle architecture have to be considered.

Based on classical knowledge the ventricular walls in the human may be thought of as three layers, the endocardial, the middle and the epicardial (Lev and Simkins, 1956). These layers can be characterized by different fiber orientations. It has also been stated that the width of the middle layer in the right ventricle is very thin (Lev and Simkins, 1956). Even though more recent investigations have revealed that the fiber orientation changes continuously along the endo-epicardial line (Streeter, 1979), in the present model study we chose a simple two-layer model of the myocardium where the layers were supposed as being at right angles. Assuming that the anisotropy

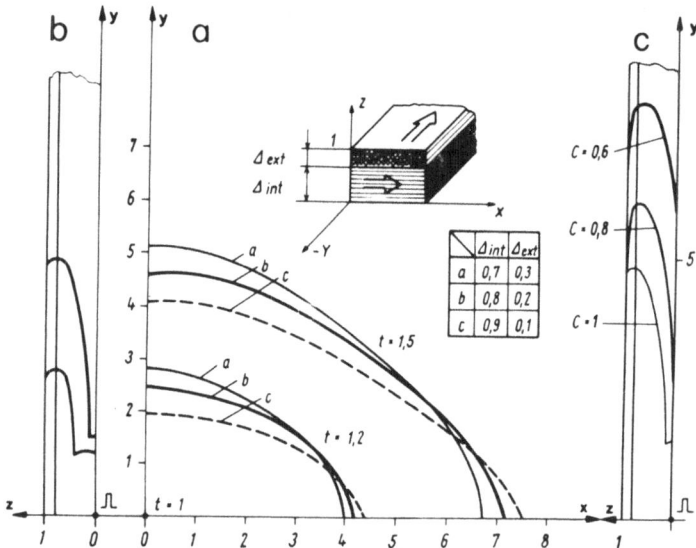

Fig. 2. Simulated activation propagation. (a) Calculated epicardial isochrones at three different internal to external layer-width ratios. Time and length are represented in relative units. The time interval from the endocardial stimulation until the epicardial breakthrough serves as a relative unit of time, the distance travelled by the activation wavefront perpendicular to the fiber axis during the time unit as a relative length unit. (b) Activation wavefronts in the zy plane corresponding to b) isochrones. (c) Activation wavefronts in the zy plane at different c velocity ratios of the internal to external layers.

of the excitation propagation is primarily associated with the arrangement of the intercalated discs found at the interfaces of adjacent cells, we took the ratio of velocities along and perpendicular
to the fiber axis to be 7 (Streeter, 1979).

The propagation of activation was treated on the basis of Huygens' principle with the condition that the elementary wavefronts are
ellipsoids.

An illustration of the results obtained is depicted in Fig. 2.
Here, in the middle, the calculated epicardial isochrones are visible
at three different internal to external layer-width ratios. A comparison of these isochrones with those obtained earlier by Préda and
d'Alché (1977) on a dog's heart shows that 0.8 and 0.2 seem to be the
best values for the thickness of the endocardial and epicardial
layers (the curves drawn with thick lines). For the calculation the
right ventricular activation data were used as we consider our two-
layer approximation as justifiable here.

⋅ On the left side of this figure the activation wavefront in the
zy plane is shown corresponding to the thick-line isochrones. In
Fig. 3 a simulated epicardial activation map is depicted on the side
of the real measured map. The simulation was made with the assumption that endocardial stimulation was applied to the three focuses
indicated. The time diagram of the operation of the stimulating
focuses was chosen in a way that a good agreement was obtained between the isochrones of the model and that in the dog experiment.

As the muscle layers of the right ventricular free wall are continued in the septum, the isochrones shown under the dotted line can
be interpreted as the representation of septal activation. The experimental data gained for the same dog in normal temperature and in
hypothermia revealed that at low temperatures the epicardial excitation is retarded but at the same time the general pattern of isochrones remained the same. In contrast, significant changes were
observed in the time-course of the body surface potential maps (d'
Alché and Préda, 1977).

A possible explanation of this phenomenon might be, that during
the cooling of the dog's body a significant temperature gradient was
generated between the epicardial and endocardial layer. If this gradient manifests itself in a non-uniform decrease of conduction velocities a significant change in the internal wavefront geometry may
develop even in the case of a similar epicardial isochrone pattern,
as it is to be seen in the diagram on the right side of Fig. 3.
(here the different c values represent the different velocity ratios
of the internal to external layers).

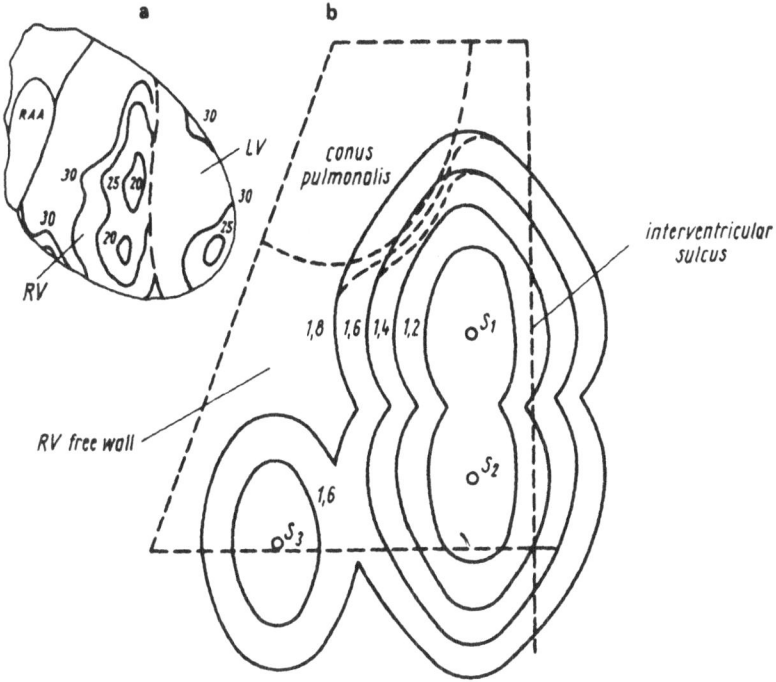

Fig. 3. Measured (a) and simulated (b) epicardial activation maps.
S_1, S_2 and S_3: focuses of endocardial stimulation.

CONCLUSION

On the basis of the averaged body surface maps corresponding to
a homogeneous group of normal and to a group of cardiac patients
suffering from different forms of conduction disturbances, we estimated the stimulus-response functions of the individual bundle
branches and their main fascicles. The knowledge of these functions
may offer a new tool for potential map interpretation.

With the help of a simple two-layer anisotropic model of the
right ventricular wall some properties of the intramural excitation
propagation were investigated. On the basis of the results a possible interpretation of the experimentally observed activation isochrones and body surface maps were outlined.

REFERENCES

d'Alché, P., and Préda, I., 1977, Distribution of cardiac thoracic
        potentials on the dog in hypothermia, Adv. Cardiol., 19:77-83,
        Karger, Basel.

Baruffi, S., Spaggiari, S., Stilli, D., Musso, E., Taccardi, B.,
    1978, The importance of fiber orientation in determining the
    features of cardiac electric field, in: "Modern Electrocar-
    diology," Z. Antalóczy, ed., Excerpta Medica, Amsterdam, 89-
    92.

Lev, M., and Simkins, C. S., 1956, Architecture of the human ven-
    tricular myocardium, Laboratory investigation, 5:396-409.

van Oosterom, A., 1978, "Cardiac potential distribution," Ph. D.
    Thesis, Amsterdam.

Préda, I., and d'Alché, P., 1977, On the temperature dependence of
    dogs cardiac activation, Adv. Cardiol., 19:33-37, Karger,
    Basel.

Streeter, D., Jr., 1979, Gross morphology and fiber geometry of the
    heart, in: "Handbook of Physiology, Section 2: The cardiovas-
    cular system," Ch. 4, American Physiological Society, Beth-
    esda.

Taccardi, B., Musso, E., de Ambroggi, L., 1972, Current and poten-
    tial distribution around an isolated dog heart. Proceedings
    of the Satellite Symposium of the 25th International Congress
    on Physiological sciences (The electrical field of the heart)
    and the 12th Colloquium Vectorcardiographicum, P. Rijlant, ed.,
    Presses Académiques Européennes, Brussels, pp. 566-572.

EPICARDIAL ST-SEGMENT MAPPING IN ACUTE MYOCARDIAL ISCHEMIA. EXAMPLES OF COINCIDENTAL EXPERIMENTAL INTERVENTIONS WHICH MAY AFFECT INTERPRETATION

Waldemar J. Wajsczuk, Jacek Przybylski, Grzegorz Sedek, Ryszard Jacek Zochowski, Tadeusz Palko, Albert Whitty, and Melvyn Rubenfire

Section of Cardiovascular Diseases Department of Medicine, Department of Research, Sinai Hospital of Detroit and Department of Medicine, Wayne State University, Detroit, Michigan, U.S.A.

The technique of multi-electrode epicardial recording of electrical activity of the heart is now commonly used for evaluation of the ischemic injury in experimental studies (Maroko et al., 1971). Its ability to follow the dynamic changes and ease of application made this method attractive and widely used in experiments involving studies of the evolution of myocardial infarction and of the influence of various interventions. Multi-electrode recording of precordial electrical signals was subsequently introduced for clinical studies (Maroko et al., 1972; Muller et al., 1975).

Experimental studies of Maroko et al. (1971, 1973, 1975, 1976) demonstrated the correlation between the magnitude of ST segment elevation and morphologic and biochemical alterations in myocardial tissue samples underlying the electrodes. The authors suggested that the magnitude of initial ST segment elevation at 15 minutes after coronary artery occlusion would predict the progression of severity and extent of myocardial damage and necrosis at 24 hours after the onset of ischemia. Subsequently, numerous publications generally interpreted the decrease in the magnitude of ischemic ST segment elevation after hemodynamic or pharmacologic interventions as indicative of a beneficial response and as an expression of the decrease of

Supported by NIH Grant HL-13737-05, Michigan Heart Association, Sinai Hospital General Research Support Fund and Cardiology Research and Education Fund and a grant from the Upjohn Company.

Address for reprints: Waldemar J. Wajszczuk, M.D., Sinai Hospital of Detroit, 6767 West Outer Drive, Detroit, Michigan 48235, U.S.A.

ischemic damage (Braunwald and Maroko, 1973; Maroko et al., 1975; Awan et al., 1976; Madias et al., 1975; Madias et al., 1976). Our observations during studies on the natural course of dynamic changes in the experimental electrograms in sustained myocardial ischemia (in experiments designed to evaluate the effects of interventions) indicated that on occasion these relations are not true (Wajszczuk et al., 1979; Zochowski et al., 1979; Sedek et al., 1975; Zochowski et al., 1977). The purpose of this report is to present examples of interventions which were found to affect the ST segments without necessarily benefitting the ischemic myocardium and which may modify the ST segment response to ischemia (and/or interventions). They may lead to difficulties and potential pitfalls in interpretation of the ST segment behaviour and to erroneous conclusions, if not recognized, thus limiting the value of epicardial ST segment mapping technique.

METHOD

Experimental observations were made in adult healthy mongrel dogs of both sexes weighing 25 - 35 kg. Under anesthesia with intra-venous sodium pentobarbital 35 mg/kg and premedication with succynyl-cholin, the animals were intubated and artificial respiration initiated (Harvard respirator, Model 607 Pump). The chest was opened with a median sternotomy incision, the pericardium incised and the heart was loosely suspended in a pericardial cradle. The anterior descending (ventral) coronary artery, its apical branch and the left circumflex coronary artery were locally dissected from the connective tissue and prepared for subsequent occlusions by tightening of a tourniquet. Four dacron strips, each containing 5 silver electrodes, approximately 1.5 cm apart, were sutured loosely at their ends to the pericardial surface in parallel arrangement. Unipolar leads were used for recordings. Multi-channel recorder (Hewlett-Packard 7798A) allowed simultaneous recording from each electrode strip in a vertical array. Switching to the next set of electrodes was done automatically at approximately 10 second intervals. The ST segment elevation was measured from the point on the ST segment 100 msec from the beginning of the QRS. Aortic pressure was monitored through a catheter introduced to the ascending aorta via the carotid artery and connected to the pressure transducer (Statham P23dB No. 8718). Animals with stable blood pressure and without arrhythmias were included for analysis.

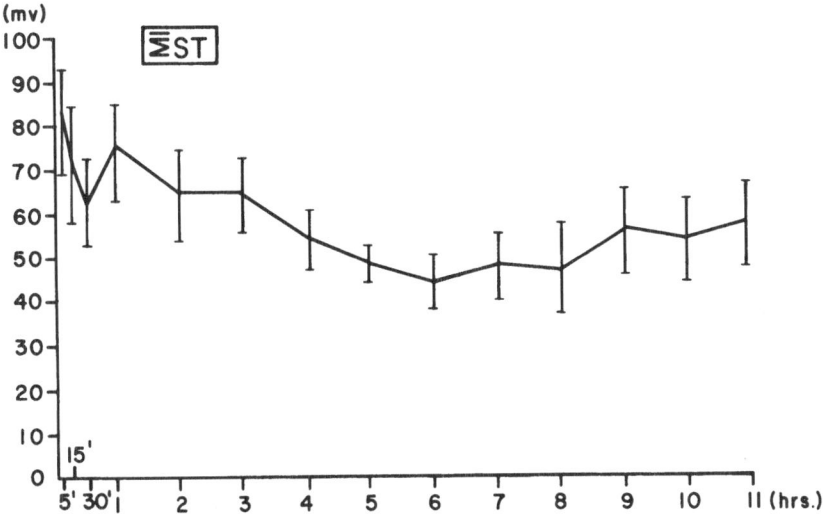

Fig. 1. Spontaneous changes (natural history) of ST segment eleva-
         tions during 11 hours of acute ischemia (mean values (ΣST)
         in a group of 10 dogs.

RESULTS

A. Natural History of Ischemia

     Natural history of ischemic ST segment elevations. Figure 1 il-
lustrates the spontaneous course (natural history) of changes of ischemic ST segment elevations. The behaviour of ST segments was
studied in a group of 10 dogs during the first 11 hours of ischemia.
The changes are expressed as a mean value of the sum of ST elevations
(ΣST) in the ischemic zones in individual animals averaged for the
whole group. The maximum is observed during the first 5 - 15 minutes
then is followed by a slight spontaneous decrease and a second maximum between 30 and 60 minutes. ST segment elevation remains fairly
stable until the third hour of ischemia. From that time on, a spontaneous gradual decrease is observed. After six hours and 8 hours,
the ST elevation is reduced by approximately 40%. A recurrent rise
is observed after 8 hours. After 11 hours, the ST elevation is decreased by only 25% in comparison with its initial maximum.

     Spontaneous variability and dynamic changes. Observations were
made in 20 dogs. Various configurations and magnitudes of ischemic
ST segment elevations were observed under different electrodes, depending on their localization in relation to the center of the ischemic area. Figure 2 shows examples of early ischemic ST segment
elevations recorded simultaneously in the same dog from different
electrode locations after 5 minutes of ischemia. Transitions between

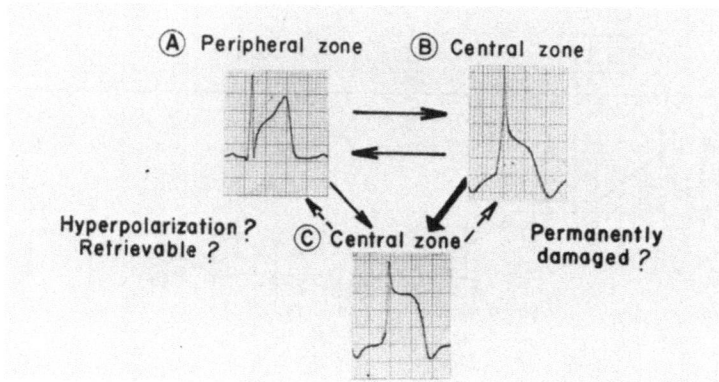

Fig. 2. Various configurations of ST segment elevations observed
early during ischemia in the periphery and in the central
zone. Recordings were obtained with unipolar epicardial
leads. Subsequent transitions between various configurations and their relative frequency are indicated by different thickness of arrows. Broken arrows indicate infre-
quent transitions.

various configurations of ST segment elevations which were commonly
seen later during ischemia are indicated by the arrows. In association with the changes in ST segments, there were alterations in the
configuration of the R waves and, in particular, changes in their
voltage in relation to the preceding baseline.

B. Intrinsic Factors, Experimental Influences and Modifications

Effects of various factors related to coronary circulation,
extension of ischemic zone, generalized hypoxia (or asphyxia) and
effects of alternating episodes of ischemia and reperfusion were
studied in separate groups of dogs. These influences can either
occur spontaneously or be inadvertantly introduced, if the experiment is not carefully supervised.

Extension of the ischemic zone. The effects on ST segment of
extending the area of ischemia were studied in 5 dogs. Examples are
illustrated in Figure 3. The electrode site under study was in the
center of the initially ischemic zone. In this experiment, as expected, first ligation of the apical branch of the anterior descending coronary artery resulted in ST segment elevation. Tightening of
the second ligature which was localized higher on the anterior descending coronary artery (Figure 3A) or on the circumflex branch
(Figure 3B) produced almost immediate decrease of ST segment elevation. This decrease persisted for the duration of maintaining the
second ligature (which was usually for a few minutes). ST segment
elevation increased again after releasing the second ligature. This

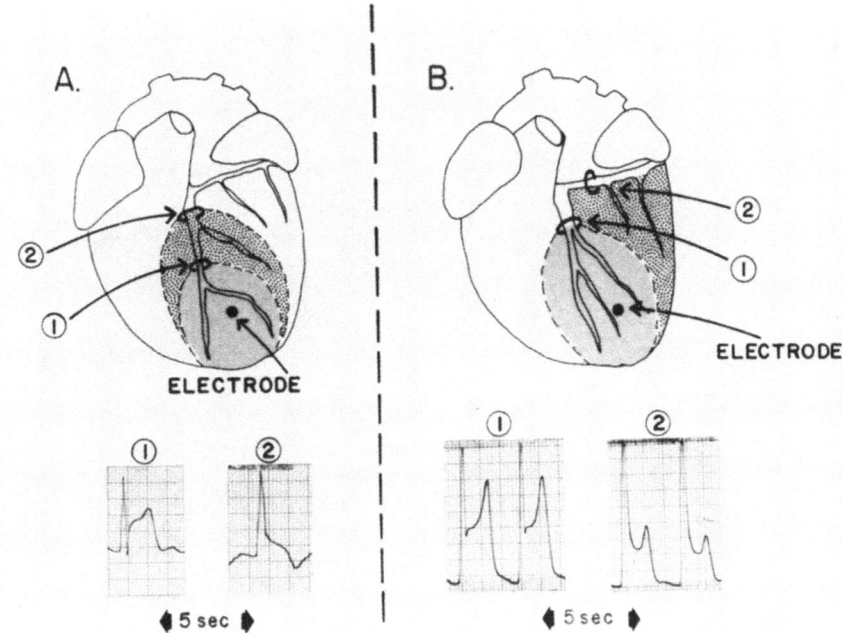

Fig. 3.   Effect of extension of the ischemic zone on ST segment ele-
          vations.   In Panel A, a second ligature was placed higher
          on the left anterior descending coronary artery.   In Panel
          B, a second ligature was placed on the circumflex branch.
          In both instances, marked decrease of ST segment elevations
          was seen within 5 seconds after placing the second ligature.

phenomenon was observed to be reproducible and could be repeated
several times in the same experiment with identical results.

   Distribution of ST elevations in the ischemic zone.   Distribution of the electrode sites within the ischemic area in regard to the
magnitude of ST segment elevations and to the extention of the ischemic zone was studied in a series of 6 dogs.   In these experiments,
a flexible cottonwick electrode moistened with Ringer's solution was
slowly swept across the ischemic area while continuously recording
the unipolar epicardial electrograms.   The different magnitudes of ST
segment elevation in relation to the position of the sweeping electrode in the periphery or in the center of the ischemic zone are represented in the graph in Figure 4.   Diagrams illustrate the locations of the ligatures.

   In experiments illustrated in Figure 4A (diagram and plot A) a
single ligature was tightened on the left anterior descending coronary artery.   It resulted in lowest ST segment elevations near the
borders of the ischemic zone and highest ST elevation in its center
(plot A in the graph).

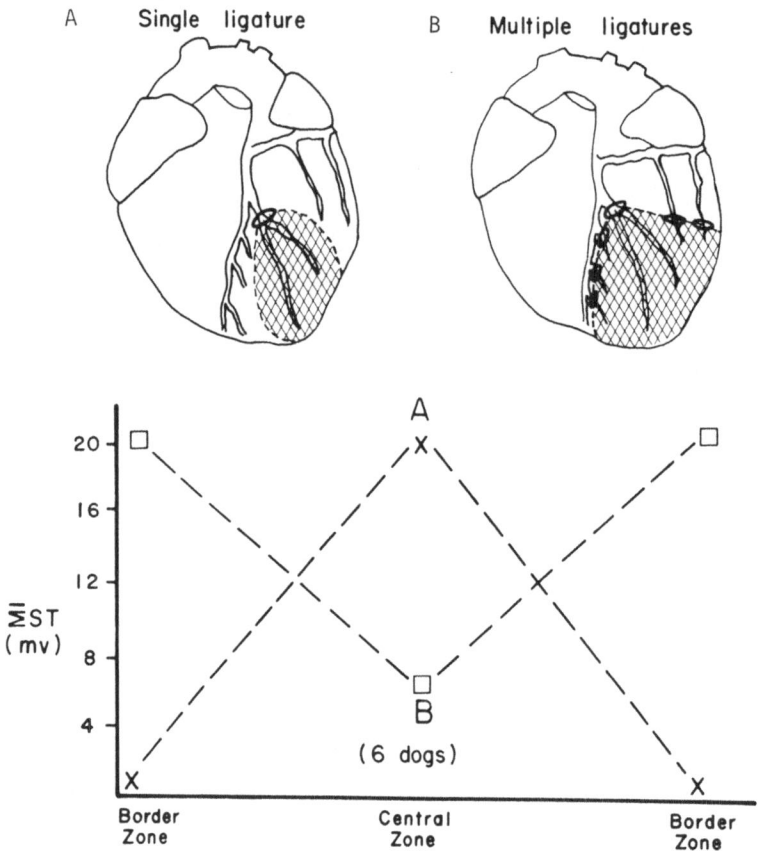

Fig. 4. Relationships between the magnitude of ischemic ST segment
        elevations and the locations of recording electrodes in the
        ischemic zone.   In Panel A, single ligature was applied.
        Lowest ST segment elevations were seen in the periphery and
        highest in the center (plot A in the lower part of the il-
        lustration).   In Panel B, multiple ligatures were applied on
        the terminal segments of small branches supplying the border
        zone of the initially ischemic area.
        Highest ST segment elevations were seen in the border zone
        and lowest (markedly decreased from those seen after place-
        ment of the original ligature) in the center of the ischemic
        zone (plot B in the graph).   (See text for details.)

    Effects of multiple ligatures.   In experiments illustrated in
Figure 4B (diagram and plot B), after initial recording of the ST
segment responses to the single ligature on the LAD branch, multiple
ligatures were placed on the terminal portions of all coronary branches (identified by visual inspection) near the borders of distribution of the LAD branch which was originally ligated.   The possible

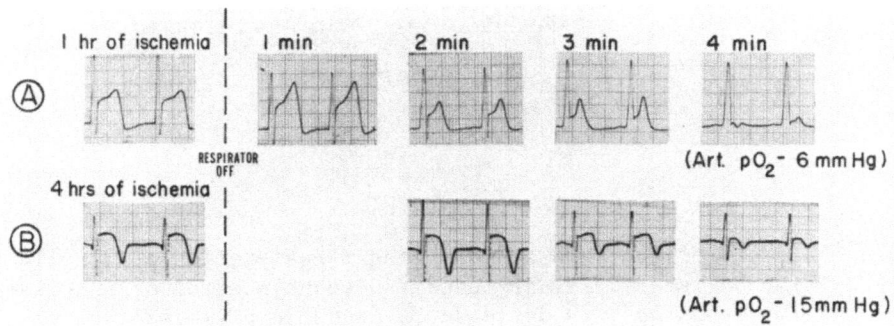

Fig. 5.   Effect of asphyxia (by discontinuation of mechanical ven-
          tilation) on ischemic ST segment elevations. There is near
          normalization of the ST segments without associated sig-
          nificant changes in the QRS complexes after 4 minutes of
          asphyxia. Examples from 2 dogs (A - after 1 hour of is-
          chemia and B - after 4 hours of ischemia). Arterial blood
          PO2 values are indicated.

sources of collateral circulation (except deep intra-myocardial
branches (Burch et al., 1970)) were thus interrupted. The resultant
ischemic area was slightly larger than the original one. A sweep of
epicardial electrodes across the new ischemic area identified highest
ST segment elevations in the border zone and lowest ST segment elevations (definitely ischemic in comparison with the pre-ischemic
control recordings) in the center of the new ischemic area (plot B in
the graph). The distribution of ST elevation was different and opposite in character to that seen in the experiments in which a single
ligature was employed. This difference could possibly reflect the
contribution of collateral circulation and of its effect on the behaviour of ST segments in response to ischemia.

    Hypoxia and asphyxia. Generalized hypoxia of the myocardium was
produced by turning off the respirator at the peak of inspiration in
8 animals with transsected vagal nerves and localized myocardial ischemia produced previously by ligation of the coronary artery branch.
Included in this study were only animals in whom there was no rhythm
disturbance and no abnormality of atrioventricular conduction during
the period of asphyxia. Figure 5 shows the examples of changes in
ST segment elevations resulting from progressing hypoxia and asphyxia
and recorded from the electrode localized near the center of the ischemic zone. In the animal in panel A, asphyxia was induced one hour
after ligation of the left anterior descending artery and in the
animal in panel B after four hours of localized myocardial ischemia.
In both examples, there is a gradual decrease of ischemic ST segment
elevation with almost complete normalization of the ST segment within
4 minutes after the onset of asphyxia. Only minimal alterations are
seen in the QRS complexes. Similar phenomenon was seen under most of
the electrodes localized in the initially ischemic zone. In the ex-

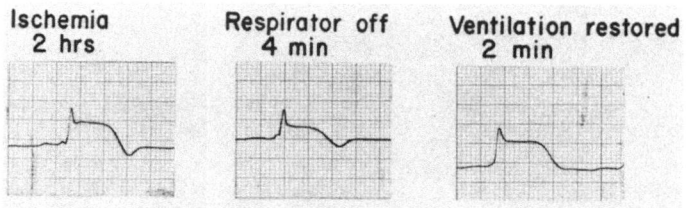

Fig. 6. Effects of asphyxia and subsequent restoration of ventila-
         tion on ischemic ST segments. Discontinuation of ventila-
         tion resulted in a decrease of ischemic ST segment eleva-
         tion. Restoration of ventilation was accompanied by recur-
         rent increase of ischemic ST segment elevation to its pre-
         vious level. Note: paper speed is 50 mm/sec.

ample from a different dog (Figure 6), reinstitution of ventilation
and restoration of acid-base balance by administration of sodium bi-
carbonate resulted in restoration of the initial electrocardiographic
pattern of acute myocardial ischemia.

Reperfusion. In 11 dogs, the coronary artery branch (LAD) was
ligated and ischemia maintained for 15 minutes. At the end of this
period, the ischemia was evaluated in regard to its severity (ΣST =
total sum of ST segment elevations recorded from the ischemic electrode points) and its extent (NST = number of electrode points recording ischemic ST segment elevation). The ligature was then re-
leased with resulting autoreperfusion of the previously ischemic
area. After 30 minutes, ligation was repeated in the same location
and again ΣST and NST were measured after 15 minutes from the same
electrode sites. Second ligation resulted in less ST segment elevation (Figure 7A) and slightly smaller extent (Figure 7B) of the ischemic zone in most of the dogs. In few of the animals, however,
increased ST elevation and larger ischemic zones were observed.

C. External Interventions-Effect of Drugs

Lidocaine. Since Lidocaine may be on occasion used during acute
experiments to control ventricular arrhythmia, its effect on ST segments was evaluated in six dogs. Lidocaine was administered by intravenous bolus in a dose of 1 mg/kg at the end of one hour of localized myocardial ischemia. Recordings were obtained after 5 minutes
when the maximum effect was observed. They revealed a marked decrease in the magnitude of ST segment elevations (ΣST) (Figure 8).
The effects persisted for 30 - 45 minutes. There was an associated
reduction in the mean arterial blood pressure of 20 mm Hg and decel-
eration of the heart rate by 15 - 25 beats/minute. There was no
effect on the extent of the ischemic zone (NST).

Steroids. Methylprednisolone was given in a dose of 30 mg/kg
intravenously by slow infusion over a 15 minute period in 5 dogs.

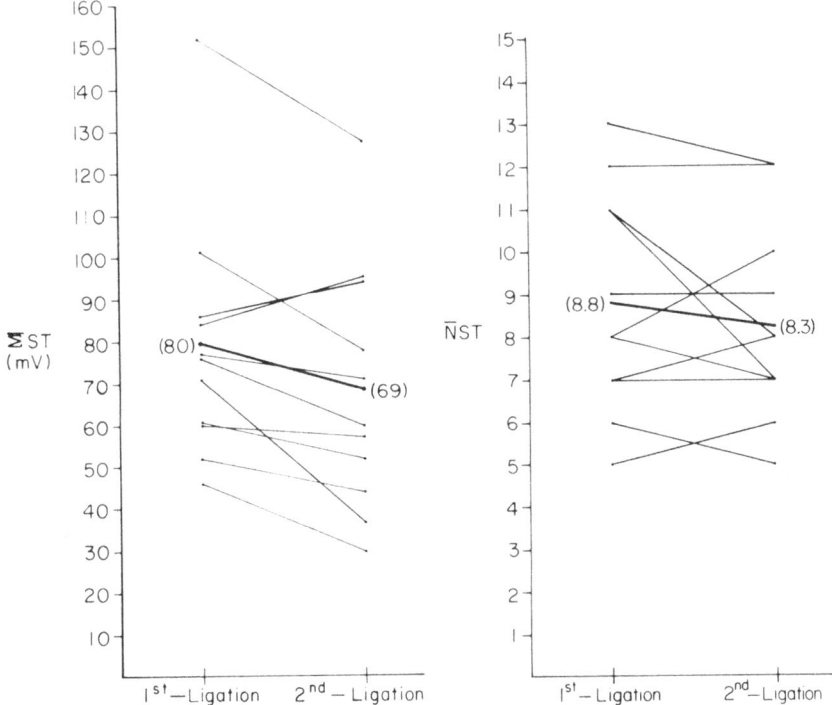

Fig. 7.   Effect of consecutive ligations separated by 30 minutes of
          autoreperfusion on the "severity" (ΣST) and extent (NST) of
          ischemia. Changes in individual dogs and mean values in a
          group of 11 dogs are presented. Reapplying ligature in the
          same location resulted in less ST segment elevations in most
          dogs. The extent of the ischemic zone after second ligation
          shows greater variation (increased or decreased).

Administration of the drug at the end of 1 hour of ischemia resulted
in a continuing decrease in the amplitude of ST elevation (Figure 9)
which was significantly different from the spontaneously occurring
decrease seen in the control group of dogs. The maximum decrease was
observed after 3.5 hours. During the subsequent 30 minutes, there
was an accelerated phase of gradual return of ST elevation to the
preintervention values. At the end of observation, six hours after
administration of the drug, there was no difference between the
groups.

DISCUSSION

     Previous extensive experimental clinical observations estab-
lished the association between the dynamic changes of the ST segments
and the onset and evolution of myocardial ischemia and damage.

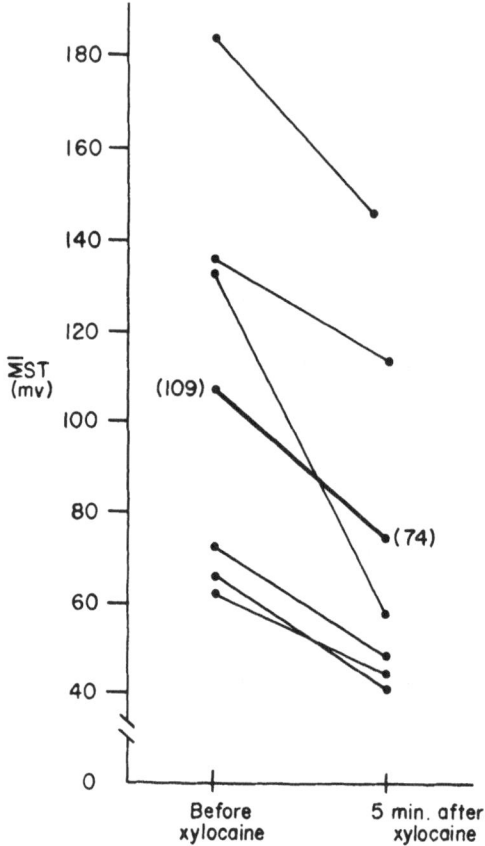

Fig. 8. Effect of intravenous infusion of Xylocaine on ischemic ST
         segment elevation. There is a uniform and significant de-
         crease of ST segment elevations in all dogs 5 minutes after
         infusion (see text for details).

Ischemic ST segment elevation is known to gradually decrease with
progression of ischemic damage and development of Q waves. In the
studies described in this paper, we have observed a spontaneous recurrent increase of ST segment elevations after 8 hours of ischemia.
While its mechanism is not well understood, it could be postulated
that it resulted from the spontaneous increase of the heart rate (by
15 - 20 beats/minute) in most of the dogs during the later hours of
experiment. Increasing tachycardia could have led to increased oxygen utilization and worsening of ischemia.

     Under certain experimental circumstances, the decrease of the
magnitude of ischemic ST segment elevation can be accepted as an expression of beneficial effects of interventions designed to limit or
counteract the ischemic damage. The outcome of these interventions

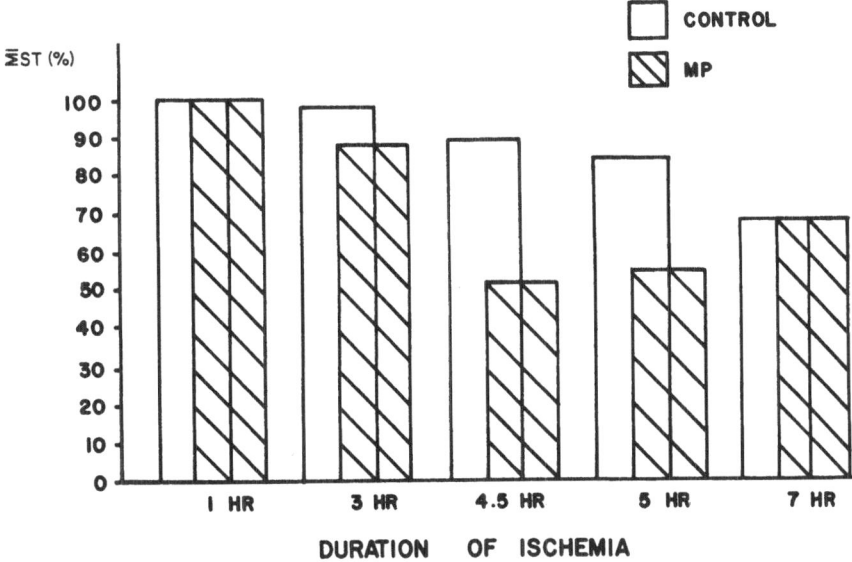

Fig. 9.   Effects of intravenous administration of Methylprednisolone
          on ischemic ST segment elevations. The effect of Methyl-
          prednisolone on ST segments in the experimental group is
          indicated by shaded bars (in percent of values after 1 hour
          of ischemia). Spontaneous changes in the magnitude of ST
          segment elevations in the control group are represented by
          unshaded bars. Maximum effect is seen in 3.5 hours after
          administration of the drug. There is no difference between
          the control and experimental groups at the end of the ob-
          servation period (6 hours after administration of the drug).
          This example illustrates a transient effect of the drug on
          ischemic ST segment elevations.

should also be evaluated in chronic experiments when the healing process is completed (Wajsczuk et al., 1977).

     There are, however, circumstances in which the magnitude of
ischemic ST segment elevation may be reduced, but this reduction may
not necessarily signify the improvement or reduction of ischemic
damage (Zochowski et al., 1979). Such circumstances may inadver-
tently develop during the course of an experiment, as a result of
intended or unintended interventions or spontaneously occurring phenomena. Their influence on the ST segments may be erroneously interpreted as a beneficial effect on the course of ischemia. Some of
these factors are reviewed in this paper, based on our observations.
They are broadly divided into the category of intrinsic factors, related to coronary circulation and potential involvement of the collateral circulation, spontaneous and induced extension of ischemic

zone, local reperfusion effect, systemic oxygenation and variability of local response to ischemia including spontaneous transitions between various forms of ST segment elevations. The other broad category includes the effects of medications which may be administered in the course of an experiment, without realizing their potential effect on ST segments (or other parameters of ischemic evaluation); for instance, Lidocaine for suppression of arrhythmias.

We have observed that the extension of the ischemic zone may produce a sudden and marked decrease in the magnitude of ischemic ST segment elevations in the center of the originally ischemic zone. This decrease does not appear to be associated with acceleration of Q wave development (which is indicative of the accelerated transition from ischemia to injury (Zochowski et al., 1979)). Also, depending on the location of ligatures and their number (and the degree of interference with or interruption of collateral circulation), entirely different distribution pattern regarding the magnitude of ischemic ST segment elevations was observed in the periphery and in the center of the ischemic zone. While with a single vessel ligature the highest ST segment elevations were seen in the center of the ischemic zone and only small elevations (or ST segment depressions) in the periphery, adding multiple ligatures encircling the ischemic zone (and most likely interrupting the collateral blood supply and obviously changing the boundary and gradients) resulted in a dramatic and sudden decrease in the severity of ischemic ST segment elevations in the center and in the appearance of severe ST segment elevations along the periphery of the ischemic zone. Whether such circumstances can occur spontaneously during the course of an experiment is unknown but such possibility cannot be disregarded entirely together with its implications.

Repeated ligations of short duration with interposed periods of reperfusion resulted, in most dogs, in less ST segment elevation and occasionally smaller extents of ischemic zone than those observed after the initial obstruction of the coronary flow. The mechanisms of those differences also are not well understood. Our previous observations (Zochowski et al., 1979) revealed that reperfusion after a sustained period of ischemia may induce a rapid decrease of ST elevation which is then, however, associated with an accelerated Q wave development.

Since no detailed studies are available, it is assumed that medications used for premedication and later for sustained anesthesia have only a continuous and steady effect on the electrocardiogram, but for instance, Atropine may affect the ischemic ST segments (Redwood et al., 1972). Similarly, antiarrhythmic medications may exert influence on the transmembrane potentials and ST segments. Lidocaine, which is usually used to suppress ventricular arrhythmias during the acute phase of ischemia, was demonstrated to modify the ischemic ST segment response (Boudoulas et al., 1978). These observations were also confirmed in our studies which revealed a decrease of the ST segment elevations which persisted for 30 - 40 minutes. The decrease in ST elevation could possibly be explained by an associated decrease in blood pressure and pulse rate resulting in a decrease in "double product" and oxygen demand in the myocardium. The example of Lidocaine further illustrates the need for careful monitoring of the effects of intended or unintended intra-experimental interventions which may affect the ST segment response and modify the responses to attempted therapeutic interventions (and thus lead to erroneous conclusions).

The effects of medications belonging to the adrenergic group on the ST segments, ischemia and injury have been well studied previously and detailed discussion of their action is not included in this review. Generally, they are known to cause acutely an increase of ST segment elevations and acceleration and extension of ischemic myocardial damage (Maroko and Braunwald, 1976).

The example of steroid administration serves to demonstrate the potentially transient effects of an intervention and again stresses the need for evaluation in chronic studies of the postulated beneficial effects of treatment intended to modify ischemia or decrease myocardial damage (Libby et al., 1973). Inadequate length of observation and follow-up of the effects of therapeutic interventions may also lead to erroneous interpretation of results and pitfalls in conclusions.

In conclusion, this brief communication was intended to review and amplify the importance of intrinsic or extrinsic factors and influences which may modify the response of the ST segments to ischemia. Since the behaviour of the ischemic ST segments is frequently used for assessment of the beneficial effects of various interventions, a thorough knowledge of their spontaneous behaviour and of external factors which may influence them and affect their spontaneous behaviour is necessary. This knowledge will assure correct interpretation of the results of experiments and of therapeutic interventions, and will prevent errors in conclusions.

REFERENCES

Awan, N. A., Miller, R. R., Vera, Z., DeMaria, A. N., Amsterdam, E. A., and Mason, D. T., 1976, Reduction of ST segment elevation with Infusion of Nitroprusside in Patients with Acute Myocardial Infarction, Am. J. Cardiol., 38:435.

Boudoulas, H., Karayannacos, P. E., Lewis, R. P., Kakos, G. S., Kilman, J. W., and Vasko, J. S., 1978, Potential Effect of Lidocaine on Ischemic Myocardial Injury: Experimental and Clinical Observations, J. Surgic. Res., 24:469.

Braunwald, E., and Maroko, P. R., 1973, Protection of the ischemic myocardium, Hospital Practice, May p. 61.

Burch, G. E., Wajszczuk, W. J., Cronvich, J. A., 1970, Spread of Activation in the Anterolateral Papillary Muscle of the Left Ventricle of the Dog Under Normal and Pathologic Conditions, 1970, Am. Heart J., 79: 769.

Libby, P., Maroko, P. R., Bloor, C. M., Sobel, B. E., and Braunwald, E., 1973, Reduction of Experimental Myocardial Infarct Size by Corticosteroid Administration, J. Clin. Invest., 52:599.

Madias, J. B., Venkataraman, K., and Hood, W. D., Jr., 1975, Precordial ST segment mapping-I. Clinical Studies in the Coronary Care Unit, Circul., 52:799.

Madias, J. E., Madias, N. E., and Hood, W. D., 1976, Precordial ST segment mapping - II. Effects of oxygen inhalation on ischemic injury in patients with acute myocardial infarction, Circul., 53:411.

Maroko, P. R., Kjekshus, T. K., Sobel, B. E., Watanabe, T., Corell, J. W., and Braunwald, E., 1971, Factors influencing infarct size following experimental coronary artery occlusions, Circul., 43:67.

Maroko, P. R., Libby, P., Corell, J. W., Sobel, B. E., Ross, T., Jr., and Braunwald, E., 1972, Precordial ST segment elevation mapping: An Atraumatic Method for Assessing Alterations in the Extent of Myocardial Ischemic Injury. Am. J. Cardiol., 29: 233.

Maroko, P. R., Davidson, D. M., Libby, P., Hagan, A. D., and Braunwald, E., 1975, Effects of Hyaluronidase Administration on Myocardial Ischemic Injury in Acute Infarction, Ann. Intern. Med., 82:516.

Maroko, P. R., and Braunwald, E., 1976, Effects of Metabolic and Pharmacologic Interventions on Myocardial Infarc Size Following Coronary Occlusion, Circul., 53:162.

Muller, J. E., Maroko, P. R., and Braunwald, E., 1975, Evaluation of Precordial Electrocardiographic Mapping as a Means of Assessing Changes in Myocardial Ischemic Injury, Circul., 52:16.

Redwood, D. R., Smith, E. R., Epstein, S. E., 1972, Coronary Artery Occlusion in the Conscious Dog. Effects of Alterations in Heart Rate and Arterial Pressure on the Degree of Myocardial Ischemia, Circul., 46:323.

Sedek, G. S., Zochowski, R. J., Wajszczuk, W. J., Whitty, A. J., Kiso, I., Freed, P. S., Moskowitz, M. S., Kantrowitz, A., and Rubenfire, M., 1975, Demonstration of lack of persistence of effectiveness of intra-aortic balloon pumping of short duration in acute myocardial ischemia, Trans. Amer. Soc. Artif. Internal Organs, 21:555.

Wajszczuk, W. J., Przybylski, J., Zochowski, R. J., Elfont, E. A., Roszka, J. P., and Rubenfire, M., 1979, Natural History of Acute Myocardial Ischemia: Electrocardiographic Epicardial Mapping and Nitroblue Tetrazolium Studies, in: "Progress in Electrocardiology," P. W. Macfarlane, ed., Pitman Medical, Tunbridge Wells, 220-225.

Wajszczuk, W. J., Zochowski, R. J., Sedek, G., Elfont, E. A., Cascade
    P., Roszka, J., Przybylski, J., Rubenfire, M., and Kantrowitz,
    A., 1977, Experimental demonstration of the ability of intra-
    aortic balloon pumping to reduce the infarct size, Am. J.
    Cardiol., 39:259.
Zochowski, R. J., Wajszczuk, W. J., Sedek, G. S., Elfont, E. A.,
    Roszka, J. P., and Rubenfire, M., 1979, Reduction of Adverse
    Effects of Post-Ischemic Reperfusion by Intra-Aortic Balloon
    Pumping: Electrocardiographic Epicardial Mapping and Nitro-
    blue Tetrazolium Studies, in: "Progress in Electrocardiology,"
    P. W. Macfarlane, ed., Pitman Medical, Tunbridge Wells, 473-
    478.
Zochowski, R. J., Wajszsczuk, W. J., Przybylski, J., Sedek, G. S.,
    Kantrowitz, A., and Rubenfire, M., 1977, Intra-aortic balloon
    pumping in myocardial ischemia: The effect of pumping dura-
    tion and delay, Trans. Amer. Soc. Artif. Internal Organs,
    23:95.

NATURAL HISTORY OF EXPERIMENTAL MYOCARDIAL ISCHEMIA.

OBSERVATIONS IN ACUTE AND CHRONIC STUDIES

Waldemar J. Wajszczuk, Ryszard Jacek Zochowski,
Jacek Przybylski, Nicholas Z. Kerin, and Melvyn
Rubenfire

Section of Cardiovascular Diseases Department of
Medicine, Sinai Hospital of Detroit and Department
of Medicine, Wayne State University, Detroit,
Michigan, U.S.A.

Introduction of ST segment mapping to experimental (Maroko et al., 1971) and clinical investigations (Awan et al., 1976; Madias et al., 1975 and 1976) on the effects of interventions intended to reduce ischemic myocardial damage was followed by a renewed interest in studying the effects of ischemia on the electrocardiogram. Although the effects of ischemia on the endocardial, intramural, epicardial and precordial electrocardiograms have been the subject of studies for several decades (Madias, 1979), there is relatively little knowledge regarding the natural history of these changes based on continuous observations over extended periods of time.

Some of the studies attempted to correlate the electrocardiographic changes observed early after the onset of ischemia with those after 24 hours for prediction of the ultimate size of the myocardial infarction (Selwyn et al., 1978; Braunwald and Maroko (1976)). Similarly, electrocardiographic changes and myocardial enzyme loss during the acute phase of ischemia were compared with electrocardiographic changes and infarct size after one week (Ginks et al., 1972).

Supported by NIH Grant HL 13737-05, Michigan Heart Association, Sinai Hospital General Research Support Fund and Cardiology Research and Education Fund.

Address for reprints: Waldemar J. Wajszczuk, M.D., Sinai Hospital of Detroit, 6767 West Outer Drive, Detroit, Michigan 48235, U.S.A.

Addition of parameters derived from measurements of the R and Q waves
to studies on ST segment evolution, improved the precision of experimental observations (Selwyn and Shillingford, 1977).

We have previously reported on continuous observations of electrocardiographic changes occurring in epicardial mapping during 11
hours of ischemia (Wajszczuk et al. 1977). The subject of this study
was to compare the changes observed in epicardial electrocardiograms
in the same animals during the early phase of ischemia (after 1 hour)
with those seen in chronic phase of experiments (after 6 weeks).

METHODS

Myocardial ischemia was produced in 10 mongrel dogs of both
sexes by ligating the left anterior descending (ventral) coronary
branch supplying the anterior wall and apex of the heart. Dogs were
anesthetized with intravenous sodium pentobarbital (25 mg/kg) and
artificially ventilated. Central aortic pressure and standard
electrocardiogram were monitored. The animals who developed hypo-
tension exceeding 25% of the control value or arrhythmias requiring
treatment were rejected from the study.

Epicardial mapping was obtained using a multi-channel recorder
(Hewlett Packard 7798A) and 20 electrodes arranged in four rows of
five electrodes mounted approximately 15 mm apart in strips of dacron
mesh. Strips were loosely sutured at their ends to the epicardium to
prevent tension injury and special caution was undertaken to avoid
pressure effect.

The magnitude of ST segment elevations and the height of the R
and depth of the Q waves were measured from individual electrode
points in millivolts and the values for each parameter were added up
and expressed as a sum for each animal (ΣST, ΣR and ΣQ). The mean
values for the group of animals and standard error of the mean (SEM)
were calculated at predetermined intervals. An electrode point was
considered to record ischemic changes if ST segment elevation 100
msec after the beginning of the QRS exceeded the control measurement
by 1.5 mV. The ST segment depressions were regarded as zero value.
The R and Q wave voltages were measured between the isoelectric line,
prior to the QRS complex and the peak of their respective deflection.
Alterations in the R waves were calculated later as the percentage
change of the R wave voltage (ΔR). The extent of the ischemic zones
was determined by the number of electrode points which recorded ischemic ST segment elevations (NST). The extent of the infarcted
zones was determined by the number of electrodes which recorded new
Q waves (NQ).

The chest was opened through the fifth left intercostal space
under sterile conditions. During the preparatory procedures, the

epicardial surface of the heart was maintained moist with a saline
solution, pre-warmed to body temperature. After completion of the
preparatory procedures, the initial epicardial ECG map was obtained
and the ligature was then tightened. The chest wound edges were approximated with loose sutures. Subsequently, epicardial mapping was
performed at 5, 15, 30 and 60 minutes after induction of ischemia.
The electrode strips were then removed but the sutures affixing the
strips to the epicardial surface were left in place for future site
identification. Loose pericardial sutures were employed to prevent
tamponade and the chest was permanently closed.

After completion of the acute phase of the experiments, IV
fluids and pleural drainage were continued for 24 hours. An anti-
biotic was administered (Keflin 1 gram/24 hours IM) for one week.
After initial recovery from anesthesia, the dogs were allowed to move
freely in kennels where they remained for the subsequent six weeks.

At the end of the six week period, the animals were anesthetized, the chest was opened with a midsternal incision and the heart
again suspended in the pericardial cradle. The epicardial maps were
obtained from the original electrode locations by identifying old
suture markers. In addition, positioning of the electrode strips
was verified with photographs taken during acute experiments. After
completion of mapping, the animals were sacrificed.

RESULTS

Epicardial maps obtained in the same dogs during the chronic
phase of experiments were compared with those obtained during the
acute phase of the experiments in regard to the extent and "severity"
of ischemia (ST elevation) and infarction (Q waves and R wave loss).

Extent of the Infarcted Zone

The extent of the zone of infarction after 6 weeks was measured
based on the number of electrode locations which recorded new Q waves
(NQ). Comparison with mapping during the acute phase of the experiments (after 60 minutes of ischemia) indicated that only 60% of the
electrode locations (NST) which initially showed ischemic ST segment
elevations developed Q waves after 6 weeks (Figure 1).

Extent of the Ischemic Zone

Epicardial mapping after six weeks indicated the presence of a
zone with persistent ischemic ST segment elevations (NST). This zone
comprised approximately 31% of the extent of the ischemic zone observed after 60 minutes of ischemia (Figure 1).

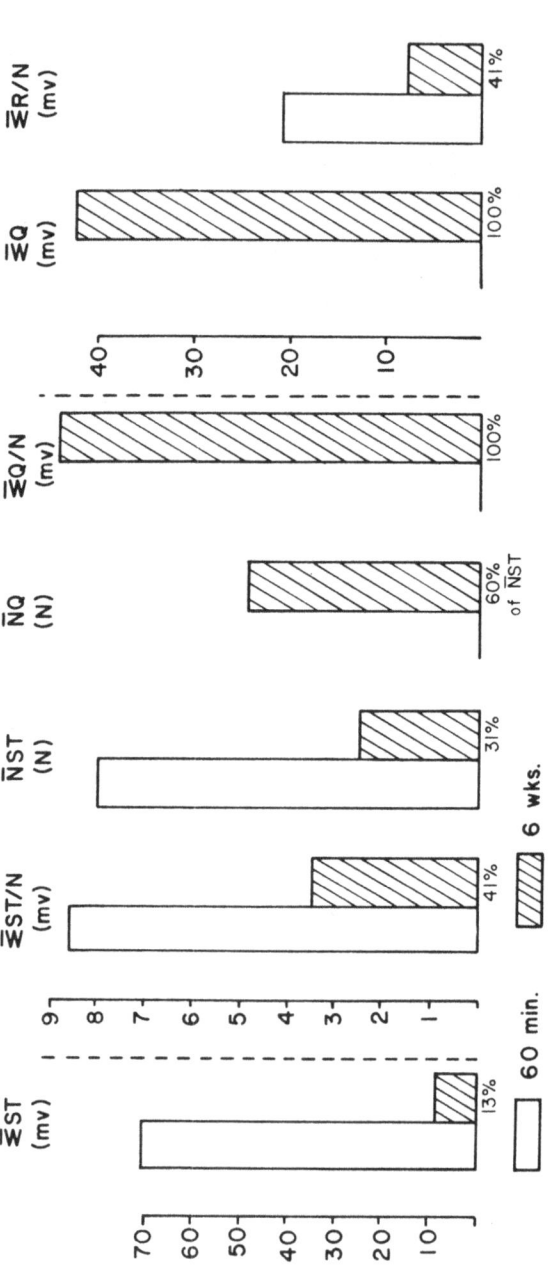

Fig. 1. Comparison between the parameters of epicardial electrocardiographic mapping recorded in the same experimental animals during the acute phase of myocardial ischemia (after 60 minutes) and during the chronic phase of infarction (after 6 weeks).

Total Extent of the Ischemic and Infarcted Zones

Comparison between the extent of the zone of ischemia (NST) after 60 minutes and the extent of combined zones of ischemia (NST) and infarction (NQ) after 6 weeks indicated that the extent of these combined zones was approximately 10% smaller after six weeks (Figure 2). This observation suggests that some degree of spontaneous reduction of the extent of ischemia and injury can occur in the course of development of an acute myocardial infarction.

"Severity" of Ischemia

The magnitude of ST segment elevations was interpreted as an expression of the "severity" of ischemia (ΣST - "total" severity of ischemia in the ischemic zone and ΣST/N - "local" severity of ischemia averaged per electrode point recording ischemic ST segment elevation). Mapping after six weeks revealed persistence of ST segment elevations (ΣST) which amounted to approximately 13% of ΣST after 60 minutes of ischemia (Figure 1). The magnitude of ST segment eleva-

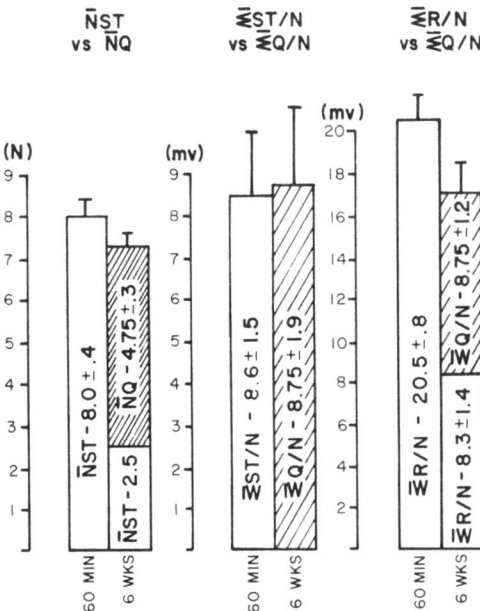

Fig. 2. Comparison between the extent of ischemia after 60 minutes and the extent of ischemia and infarction after 6 weeks. Relationship between a) ST segment elevations and Q waves and b) R and Q waves after 60 minutes and 6 weeks.

tion per electrode point which recorded ischemic ST elevation (ΣST/N) was markedly reduced after six weeks to approximately 41% of that seen after 60 minutes (Figure 1).

"Severity" of Myocardial Injury/Infarction

The magnitude of developing Q waves and the degree of R wave voltage loss were interpreted as an expression of the severity of myocardial damage resulting from ischemia.

No comparison could be made between the ΣQ or ΣQ/N values after six weeks and after 60 minutes of ischemia since precise measurements of Q waves could not be made after 60 minutes of ischemia due to their later development. It can be accepted that these parameters (ΣQ and ΣQ/N) measured after six weeks represent the final result of myocardial damage and infarction (Figure 1).

R Waves

For the purpose of this study, the R wave voltage loss was calculated at the electrode locations which showed ischemic ST segment elevations during the acute phase of experiments (after 60 minutes of ischemia). Marked R wave voltage loss (which amounted to approximately 60% R wave voltage reduction per electrode location - ΣR/N) was observed after six weeks (Figure 1).

There was no significant change of the R wave voltage outside the ischemic zone.

It was also of interest to note that the sum of the persistent R wave and of the new Q wave voltage after six weeks approximated the value of R wave voltage observed initially after 60 minutes of ischemia, at the same electrode locations (Figure 2). The significance of this observation is undetermined at present and cannot be explained based on our current knowledge and understanding (See Discussion).

Transition from Ischemic ST Segment Elevation to Q Wave Development

Although the analysis of the transitions from ischemic ST segment elevations to Q wave development and R wave voltage loss in individual electrode locations or in individual experiments did not follow any uniform patterns or show direct relationships (Wajszczuk and Przybylski, unpublished observations), comparison of the mean numerical values of local ST segment elevations per electrode point (ΣST/N) with the mean values of magnitude of the Q waves which developed after six weeks showed a similarity (Fig. 2) (See Discussion).

DISCUSSION AND CONCLUSIONS

The studies described in this paper encompass observations on epicardial electrocardiographic mappings performed during the acute phase of myocardial ischemia and during the chronic phase of experiments in the same animals after six weeks when myocardial healing is completed and myocardial infarction is fully developed and presumably stable. No further interventions were employed after the infarct had been produced by the ligation of the coronary artery branch and the dogs were maintained with a routine support which would allow their successful survival. They were allowed to ambulate depending on their physical condition. The only limitation was that imposed by the kennels. The reproducibility of epicardial mapping was assured by suture markers on the epicardium and verifying the positioning of the electrodes during the chronic phase of experiments by comparison with photographs taken during the acute phase.

Observations af er six weeks revealed the presence of a relatively large zone with persistent ST elevation which approximated 30% of the extent of the ischemic zone observed initially after 60 minutes of ischemia. There was, however, a small decrease of approximately 10% of the extent of combined zones of ischemia (with persistent ST segment elevations) and infarction (with Q waves). Most importantly, only 60% of the electrode locations which showed acute ischemic changes at 60 minutes revealed Q wave development at the epicardial surface (Holland and Brooks, 1975). This decrease could be due to salvage of the peripheral zone of ischemia due to the development and contribution of the collateral circulation to the perfusion of the borders of the initially ischemic zone.

The magnitude of ST segment elevations at the electrode locations in the periphery of infarction which recorded ST segment elevations after six weeks was significantly reduced (by 40%) in comparison with the magnitude of ST segment elevations after 60 minutes of ischemia. There was no associated Q wave development at the same electrode sites. It is unlikely that the persistent ST segment elevations were an indication of the process other than ischemia, for instance, an aneurysm, since no aneurysmatic changes were observed in these locations on direct inspection nor during gross pathology studies. It could be postulated that the reduced magnitude of ischemic ST segment elevations does indeed express the persistent but decreased severity of ischemia in these locations. Whether the myocardial cells in these locations were fully viable and able to contract, could not be ascertained from current studies; however, our previous electronmicroscopic studies (Roszka et al., 1976) revealed fairly good preservation of the ultrastructure with only minimal cellular disorganization at the electrode locations which showed persistent mild ST segment elevations without Q wave development.

There was marked reduction of the total R wave voltage in the ischemic and infarcted zone (by 60%). The magnitude of R wave voltage loss appeared to be related to the magnitude of the developing Q waves. The significance of this observation is undetermined at present and cannot be fully explained based on our current knowledge. However, our observation in dogs assisted with intra-aortic balloon pump indicated that these combined parameters can be of value during studies involving interventions intended to reduce the infarct size (Wajszczuk et al., to be published).

This comparison between the early ST elevations and resultant Q waves (Figure 2) suggest that there may be a direct relationship between the magnitude of ST segment elevations during the acute phase of ischemia and the ultimate magnitude of the Q waves after an infarct is well established. Although this relationship is not generally true for individual electrode locations or individual animals, it appears to be valid for the whole group of experimental animals. Whether this comparison will be of any predictive value, will await further confirmation in future experimental studies. Our observations in a different experimental group in which dogs with acute myocardial ischemia were supported with intra-aortic balloon pumping in an attempt to reduce the ischemic injury (Wajszczuk et al., to be published), appear to indicate that this index may be of some predictive value. In those studies, the ratio between $\Sigma ST/N$ and $\Sigma Q/N$ was markedly lower in the dogs assisted with IABP than in the control group.

In conclusion, the most important finding from this study was that, based on epicardial mapping, Q waves indicative of myocardial scarring developed in only 60% of the initially ischemic zone. The total zone of ischemia and infarction was approximately 10% smaller than the initial zone of ischemia observed after 60 minutes. There was evidence of a large zone showing persistent ischemic ST segment elevation (30% of the initially ischemic zone). Whether this zone is potentially salvageable or its presence could have ultimate detrimental effect on survival (i.e., possibility of arrhythmogenesis), deserves further study.

REFERENCES

Awan, N. A., Miller, R. R., Vera, Z., DeMaria, A. N., Amsterdam, E. A., and Mason, D. T., 1976, Reduction of ST segment elevation with infusion of Nitroprusside in patients with acute myocardial infarction, Am. J. Cardiol., 38:435.
Braunwald, E., and Maroko, P., 1976, ST Segment Mapping. Realistic and Unrealistic Expectations. Editorial, Circul. 54:529.
Ginks, W. R., Sybers, H. D., Maroko, P. R., Covell, J. W., Sobel, B. E., and Ross, J., Jr., 1972, Coronary Artery Reperfusion. II. Reduction of Myocardial Infarct Size at One week after the Coronary Occlusion, J. Clin. Invest., 51:2717.

Holland, R. P., Brooks, H., 1975, Precordial and Epicardial Surface
    Potentials During Myocardial Ischemia in the Pig. A Theore-
    tical and Experimental Analysis of the TQ and ST Segments,
    Circul. Res., 37:471.
Madias, J. B., Venkataraman, K., and Hood, W. D., Jr., 1975, Precor-
    dial ST Segment Mapping. I. Clinical Studies in the Coronary
    Care Unit. Circul., 52:799.
Madias, J. E., Madias, N. E., and Hood, W. D., Jr., 1976, Precordial
    ST Segment Mapping. II. Effects of Oxygen Inhalation on
    Ischemic Injury in Patients with Acute Myocardial Infarction,
    Circul., 53:411.
Madias, J. E., 1979, Review. Electrocardiography in Myocardial In-
    farction, J. Electrocardiol., 12(4):411.
Maroko, P. R., Kjekshus, J. K., Sobel, B. E., Watanabe, T., Covell,
    J. W., Ross, J., Jr., and Braunwald, E., 1971, Factors in-
    fluencing infarct size following experimental coronary artery
    occlusions, Circul., 43:67.
Roszka, J. P., Elfont, E. A., Kobernick, S. D., Zochowski, R. J., and
    Wajszczuk, W. J., 1976, Modification of the Ultrastructure of
    Myocardium Adjacent to Chronically Infarcted Areas by the
    Intra-Aortic Balloon Pump, Micron 7:293.
Selwyn, A. P., and Shillingford, J. P., 1977, Precordial mapping of
    Q waves and RS ratio changes in acute Myocardial Infarction,
    Cardiovasc. Res., 11:167.
Selwyn, A. P., Fox, K., Welman, E., and Shillingford, J. P., 1978,
    Natural History and Evaluation of Q waves During Acute Myo-
    cardial Infarction, Brit. Heart J., 40:383.
Wajszczuk, W. J., Przybylski, J., Zochowski, R. J., Elfont, E. A.,
    Roszka, J. P., and Rubenfire, M., 1979, Natural History of
    Acute Myocardial Ischemia: Electrocardiographic Epicardial
    Mapping and Nitroblue Tetrazolium Studies, in: "Progress in
    Electrocardiology," P. W. Macfarlane, ed., Pitman Medical,
    Tunbridge Wells, p. 220.
Wajszczuk, W. J., and Przybylski, J., Unpublished observations.
Wajszczuk, W. J., Rubenfire, M., Zochowski, R. J., Przybylski, J.,
    Elfont, E. A., Roszka, J. P., Freed, P., Kiso, I., Hamada, O.,
    and Kantrowitz, A., Protective Effect of Intra-Aortic Balloon
    Pumping on Chronic Experimental Myocardial Infarction. I.
    Electrophysiologic and Nitroblue Tetrazolium Studies. (To be
    published).

4. CLINICAL APPLICATIONS OF CARDIAC ELECTRIC

   FIELD MAPPING

AXIS CONCEPT IN BODY SURFACE MAPPING

H. Abel and G. Schoffa

St. Josephs Hospital, Wiesbaden and
Institute of Biophysics
University Karlsruhe, FRG

The axis concept is well known both in evaluation and in documentation of Ecg and Vcg for many years. In the Einthoven Ecg the
horizontal axis left to right gives the main axis, to which the
vector in the frontal plane will be referred in angle and in magnitude. In every kind of measurement of the orthogonal Ecg and in the
Vcg the parameters, which are very often used in diagnostics and in
documentation are the angles of azimuth and elevation and the spatial
magnitude of the vector. The normal values and the diagnostic criteria are very well documented in a large number of patients and col-
lectives. This holds also for the so called polarvector, the perpendicular vector of all directly measured vectors.

The body surface maps (BSM) are to be considered as pictures of
potential distributions, which are arranged in the three dimensions
of space. Therefore we suspected that angles of these potential distributions would also be of diagnostic interest and perhaps could
serve as a diagnostic criterion.

Looking for these possibilities we have developed two concepts:
1) a two dimensional concept with an angle of gradient,
2) a three dimensional concept with two angles like azimuth and elevation.

The angle of gradient is postulated to be an angle in the plane
of the map. The point of minimum will be joined with the point of
maximum by a straight line. The angle will be measured between this
and the horizontal line (Fig. 1). If there exist two maxima or
minima the lower one will be subordinated and joined rectangularly to
the straight line described above. The length of this constructed
vector is determined by the potential difference between maximum and

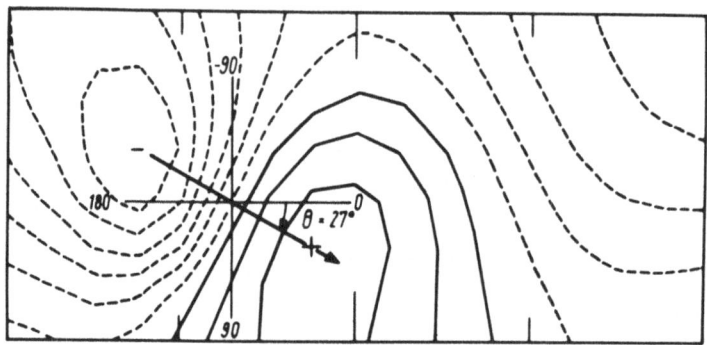

Fig. 1.   Zero is defined to be at the left side of the patient,
          downward counts +90o and upward -90o.

minimum. The length of the rectangular line including the connection point will characterize the second minimum or maximum.

It is of interest that this gradient angle represents a directional value. The manipulation of this angle and also of the angles
azimuth and elevation with statistic methods like mean values, standard deviation, variance or similar tests therefore is not correct,
though this often is neglected. Downs and Liebman (1969) accentuated
this fact several years ago. Instead of these values we used mean
directions, circular standard deviations a.so. for our statistical
calculations. These values were calculated by angle functions like
sinus and cosinus.

Healthy persons and patients with pictures of LBBB and LAH show
different gradient angles in identical time intervals. Table 1
demonstrates the different gradient angles for normal persons and for
LBBB and LAH patients. It becomes clear that this gradient angle is
a capable measure for separating these three groups of pathology.
For the statistic treatment we used the Mardia-W-test, a non para-
metric χ^2 test.

In Table 1 the calculated values are depicted in the right
column. The calculated χ^2 values are values with a $p \leq 0.005$. The
gradient angle therefore possesses a high significance in separating
the three different pathologic groups.

From the maximal amplitudes of QRS or, which means the same as
the voltage differences between minimum and maximum in the three
groups (normal, LBBB and LAH), mean directions and circular standard
deviations were calculated. The discriminatory power was fixed in a
manner described by Schoffa and Abel (1980) using a multivariate
method. This value is very significant and shows also the excellent
separation for the maximal amplitude of QRS between the groups under
investigation.

Table 2. Spherical Data in BSM

QRS Maximum

	Direction cosines			Prevalent direction			Rayleigh-Watson-Test
	1	m	n	Az.	El.	r_o	χ^2
Normal (N = 11)	.656	.602	.353	46	68	.96	30.3
LBBB (N = 7)	.868	-.249	.278	23	72	.95	18.8
LAH (N = 7)	.910	.256	.186	19	78	.96	19.5

QRS Minimum

	Direction cosines			Prevalent direction			Rayleigh-Watson-Test
	1	m	n	Az.	El.	r_o	χ^2
Normal (N = 11)	-.100	.953	.226	84	77	.98	32.0
LBBB (N = 7)	-.105	.951	.229	83	76	.98	20.3
LAH (N = 7)	.482	.699	.415	59	64	.95	18.8

Table 1. Gradient Angles in QRS

QRS segment	Normal		LBBB		LAH		Multi-sample Mardia-W-Test	
	Mean direct. °	Circular std.dev. ±	Mean direct. °	Circular std.dev. ±	Mean direct. °	Circular std.dev. ±	χ^2 observed	χ^2 0.99,dt=4
1/6	49	5.3	9	6.9	126	22.0	26.0	13.277
2/6	37	4.5	17	4.3	-20	9.5	26.0	13.277
3/6	28	6.9	16	3.2	-22	8.8	26.1	13.277
4/6	21	5.7	14	2.9	-40	9.6	18.6	13.277
5/6	5	4.5	9	5.2	-56	5.0	16.8	13.277
6/6	4	4.5	0	7.2	-78	16.0	14.6	13.277

The body surface maps are pictures of electrical events in space. The three dimensional concept therefore should be superior to the two dimensional one. For this reason we transferred the idea and the definition of the angle azimuth and elevation to the maps. Instead of the mean values we estimated the spatial prevalent direction. In Table 2 the calculated values azimuth and elevation and r_0 as the spherical mean vector are demonstrated. For $r_0=1$ all angles have the same values.

The dispersion of the two angles is caused partly by the variation in the localizations of the electrodes for different patients, but also by the variable configuration of the thorax and the different position of the heart in the chest.

For statistical treatment of the spherical directional measures the Rayleigh-Watson-χ^2-test was used as an index of quality with regard to the cluster of the angles in a group. The χ^2 values in our example are always larger than 18 for all groups. This is equivalent to a $p \leq 0.01$.

Therefore the differences of azimuth and elevation have to be considered as significant for the three groups investigated. We already had expected such results from the analysis of numerous Vcg and Ecg data and publications, which indicated such a behaviour of the axes. The advantage of our method lies in the possibility of using directly measured three dimensional values. These values can be measured in maps and are not calculated by the QRS amplitudes as in Vcg or Ecg.

REFERENCES

Downs, D., and Liebman, J., 1969, Statistical Methods for Vector-
     cardiographic Directions, IEEE Trans. BME, 16:87.
Schoffa, G., and Abel, H., 1980, Extraction of useful data in body
     surface maps, in: "7 Internat. Cong. Electrocardiol. 1980,"
     Abstracts, Lisboa, p. 19.

ON-LINE ECG MAPPING BY A SMALL MICROPROCESSOR SYSTEM

G. Schoffa

Institute of Biophysics
University Karlsruhe
Karlsruhe, FRG

Modern microprocessor techniques inaugurate new possibilities in the data acquisition and post-processing of body surface mapping.

Microprocessors
- are small in size
- are flexible for efficient programming
- have simple techniques for data transfer, and
- make it possible to construct low-priced systems.

We have constructed and built an experimental microprocessor for the hardware and software study in evaluation of the body surface maps.

HARDWARE

Figure 1 shows the block diagram of the hardware. The design uses the MOS Technology microprocessor 6502 set which has an address space of 64 K Byte installed in the desk computer system Commodore PET. The memory map is represented by

```
    0 -  1023 Bytes System
 1024 - 25600   "    Program RAM
25601 - 29952   "    Data Vector
29953 - 31126   "    Resident Assembler Utility
31127 - 32767   "    Graphic System
32768 - 36863   "    Video RAM
36864 - 49151   "    Disc System (Floppy)
49152 - 65535   "    Basic and System ROM
```

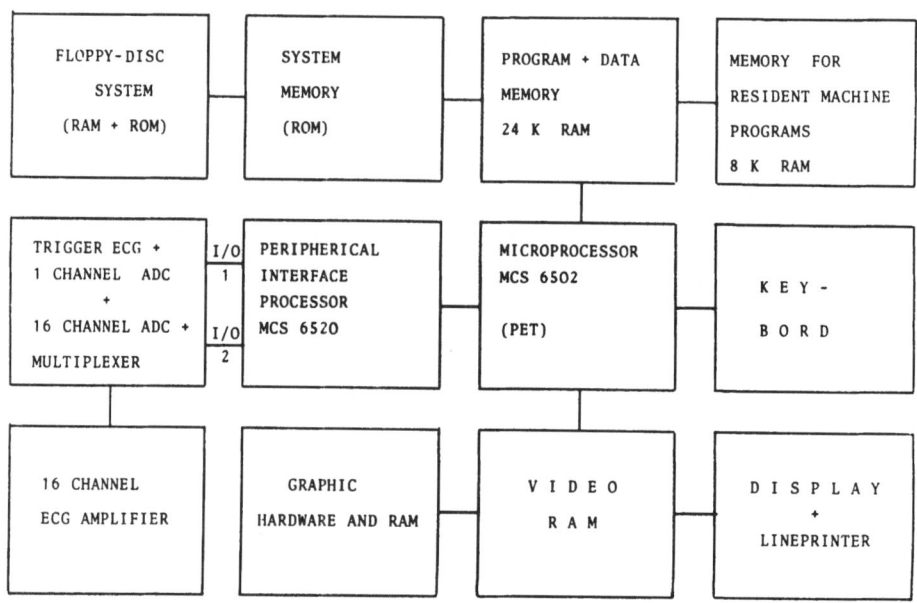

Fig. 1. Hardware schematics

The overall program control is in BASIC, but all fast functions, such as data transfer, filtering, graphical software, etc. are in the machine language and firmly resident in the memory.

ECG data derived from 128 electrodes are sent to 8 groups of 16 amplifiers. The 17th amplifier is used as a time reference. A 16 input multiplexer is switched by the software addressing the channels. A sequential free running mode is also possible, but in our case not used. All timing in the system is performed by a crystal controlled oscillator within a period of 1 microsecond. For this reason all time concerning data, such as sampling rate, time delay, etc. are realized by machine programs. The computer clock is provided as an output signal through I/O gates to permit synchronisation of the external equipment. Data from the multiplexer are fed into a 70 kHz ADC which is also strobed by a computer handshake line (CB2). The ADC input accepts input signals within 0 - 5 Volts and produces 8 bit numbers, that means an accuracy of 1:256. We use only positive levelled signals since positive values can be programmed better and more efficiently in the machine language. The sampling rate is 250 values/second. After each conversion the microprocessor makes a control for overflow, underflow, base line, and asks also for a break by the operator.

In the course of limited data storage we store only 256 values, but they are triggered by the 17th channel with its own ADC so that we then receive a complete cycle.

Analog-to-digital conversion is followed by smoothing with the best fit 2nd degree polynom approximation. Such filter programs are very slow in BASIC (50 sec for 256 values). A new evolved algorithm in machine language limits all operations to addition, subtraction and shifting of Bytes (Schoffa, in press). The smoothing time for 256 values is thereby reduced to 0.2 seconds.

Visual representation of stored ECG data serves for the examination of the results.

A machine program makes possible a 0.1 sec refresh of the video RAM and then a rapid display for the control of the data. Output characters are here "*", but the resolution of amplitudes is only 1/16, due to the 16 lines on the screen. This is good enough for the first control. High-resolution graphical display of ECG (320 x 128 pixels) requires a calculation and display time of nearly 10 sec.

The contents of the video RAM memory can be transferred to the floppy storage, and pictures can be loaded as required.

SOFTWARE FOR MAP GENERATION

An interactive software system has been developed to permit an efficient production of maps. This software system operates in conjunction with a disc operating system. Under the control of this system both segment programs and data (ECG data, video data, etc.) are stored in disc files.

The construction of isopotential maps is programmed in a high-level language as described before (Schoffa, 1978; Schoffa and Kienle, 1979), but more efficient machine programs are possible and now in preparation. A dynamic and quick representation of maps is possible by storing all maps on floppy discs, the subsequent transfer of video data in groups to 9 maps in a microprocessor memory and then a quick transfer into the video RAM with a screen representation on request.

INTERACTIVE CLASSIFICATION SYSTEM

A classification system based on the method of k-next neighbours (KNN) has been developed as an interactive system. The system performs the classification across a data base on behalf of

- learning subpopulations (normal, LBBB, LAH, etc.)
  and
- the feature vector for the case, which is to be classified.

The result is a display of the square of the Euclidean distance

- to all individual cases in a learning population,
and
- to all classes of the learning populations.

Moreover, the system displays relative weights of individual features, i.e. measured data in a new case. This table shows the significance of individual features for the classification and permits a critical overview and the improvement of the feature set.

After all, a microprocessor system is very flexible in hard- and software, and it makes possible the optimal solutions for data acquisition, and the preprocessing and postprocessing of body surface maps.

REFERENCES

Schoffa, G., (in press), Ein schneller Algorithmus zur Glättung der Kurven mit Ausgleichsparabeln, in: EDV in Medizin u. Biologie, G. Fischer, Stuttgart.

Schoffa, G., 1978, Data reduction in isopotential surface maps, in: "Modern Electrocardiology," Z. Antaloczy, ed. Excerpta Medica, Amsterdam, pp. 153-158.

Schoffa, G., and Kienle, F. A. N., 1979, Image processing in surface and gradient maps, IEEE Proceedings, Computer in Cardiology pp. 303-305.

METHODS FOR AUTOMATIC CLASSIFICATION IN BODY

SURFACE MAPPING

G. Schoffa

Institute of Biophysics
University of Karlsruhe
Karlsruhe, FRG

Our problem may be stated as follows: a digital computer receives map data for several classes of heart diseases and has to decide for a new case into which class this case should be filed. This is a well-known problem and solved in general for pattern classification: A map is described by a finite number n of variables, called features. A particular map is considered as a point in an n-dimensional pattern space encompassing the region in which patterns can occur. Sets of patterns are classified to pattern classes. Then the problem is to classify a set of unknown maps into one of the known classes.

One way is the determination of a decision boundary separating the classes. Such method requires a decision rule. The second way is the principle of "similarity" of data sets, which has to decide, whether the new pattern is "similar" to any of the known classes.

We have tested several classificators, and our criticism to map classification is the following. The first hidden problem is the dimensionality of variables. The variable space can be represented in geometrical dimensions. For the one-dimensional space the minimum of samples is 2, for the two-dimensional 4, the three-dimensional 8, and the n-dimensional 2^n. For 10 variables $2^{10} = 1024$ samples are necessary, if nothing is known about sample distribution. Foley and co-worker (Sammon et al., 1970; Kanal and Chandrasekaran, 1968) have shown that for a 60 to 70% accuracy in classifying for a two-class problem as a condition applies

$$\frac{\text{No. of samples per class}}{\text{No. of variables}} > 3./.5.$$

If this ratio is smaller it may be unable to extract the intrinsic dimensionality, which is an additional dimensionality by hidden unknown variables (Meisel, 1972).

If a normal distribution can be accepted the classification is based on mean values and their standard deviations. The accuracy of a mean value is $\sigma/\sqrt{N}$ with N as the number of samples. This implies that 100 samples are needed to estimate a mean with an error of 10 percent or smaller.

Above all, the main problem in classification of body surface maps is the small number of cases in the course of laborious techniques in data acquisition and postprocessing. To overcome this dilemma it is necessary to
1. reduce the number of variables
2. use only variables with good standard deviations, and
3. reduce the number of classes.

To reduce the number of classes means to convert a k-class problem into a series of two-class problems by successive dichotomy, that is a successive splitting of all remaining classes into two groups.

The decision boundaries resulting from successive dichotomies, however, may be less satisfactory than those resulting from the decision of all classes. Decision functions can separate the samples perfectly only if the first-level function separates one of the group perfectly and only if this process can be repeated at every level.

To minimize the number of classes we have additionally developed a "one class classifier." This is the decision between one of the classes (normal, LBBB, etc.) and the rejection by "limit exceeded."

The third possibility is to use multivariate analysis for small data sets. For small populations the t- or Student distribution is well qualified, and we have tested the T^2-Hotelling method. The quality of separation of classes is quantified by Chi-Square, and the number of cases in each class is controlled by the degree of freedom. The Hotelling T^2-classification is a parametrical method which assumes normal distribution and then defines discriminant functions by probability densities. The normal distribution in body surface data, however, is not proved. A hard limitation is also the errors of mean values.

Nonparametric methods are not based on any distribution. Especially for small data sets they are well qualified for pattern classification.

We have tested two simple nonparametric classifiers:

1. the weighted vector differences (WVD) and
2. the k-next neighbours (KNN)

The principle of weighted vector differences classification (Post-weiler, 1979) is to add a new case to the class C* to which the Euclidian distance is given by

$$d^{(c)} = \sum_{j=1}^{N} (x_j - \bar{x}^{(c)})^2$$

has to have a minimal value, that means $C^* \rightarrow \min d^{(c)}$

With respect to the distances by a weighting factor

$$d^{(c)} = \sum_{j=1}^{N} W_j^{(c)} \cdot (x_j - \bar{x}^{(c)})^2$$

The k-next neighbour classification computes the distance d_n between the new case x and all learned cases y following

$$d_n = \sum_{j=1}^{N} (x - y_j)^2$$

The next step is to arrange the distances in a sequence from the lowest to the highest. The new case belongs to the class with the majority vote.

Both last mentioned nonparametrical methods are mostly used in our group, in consequence of their high lucidity.

REFERENCES

Kanal, L. N., and Chandrasekaran, B., 1968, On Dimensionality and Sample Size in Statistical Pattern Classification, Proc. NEC Conf., pp. 2-7.
Meisel, W. S., 1972, Computer-oriented Approaches to Pattern Recognition, Acad. Press. N.Y., pp. 12-15.
Postweiler, W., 1979, Diplomarbeit, Universitat Karlsruhe.
Sammon, J. D., Foley, D., and Proctor, A., 1970, Considerations of Dimensionality versus Sample Size, IEEE Symp. Adaptive Processes, Austin, Texas.

EXERCISE ELECTROCARDIOGRAPHY AND MONITORING OF MYOCARDIAL

INFARCTION WITH A CLINICAL MAPPING SYSTEM

R. Hinsen, J. Silny, G. Rau, R. v. Essen,
W. Merx and S. Effert

Helmholtz-Institute for Biomedical Engineering
and Department of Internal Medicine I
RWTH Aachen, FRG

INTRODUCTION

Multiple thoracic leads supply more electrical information about the cardiac electric field than the standard ECG-leads (Taccardi, 1963; Kornreich, 1973; Abildskov et al., 1977). It has been shown in several studies, that this additional information can be used to increase sensitivity and specificity of the diagnosis of heart diseases, for example myocardial infarction and WPW syndrome (Flowers et al., 1976; Vincent et al., 1977; Yamada et al., 1975; de Ambroggi et al., 1976). Lead systems with up to 256 thoracic electrodes have been used to register a maximal information content. First computerized mapping systems have been developed to manage the immense data streams that are combined with the acquisition of multiple ECG signals (Cottini et al., 1972; Wyatt and Lux, 1974; Tiberghien et al., 1976). Investigations under laboratory environment conditions could be performed with these systems but the most clinical studies, especially with seriously ill patients in the coronary care units were impossible. Prior needed conditions for the widespread application of body surface mapping are small mapping systems for clinical routine use, which are self contained, mobile and easy to operate. Such systems will allow comprehensive studies with large numbers of normal and abnormal maps. This clinical experience is necessary to evaluate the diagnostic importance of the additional electrical information in the maps.

We developed a clinical mapping system and used it for monitoring ischemia and necrosis in patients with acute myocardial infarction and for detecting ischemia during stress tests in patients with suspected coronary heart diseases.

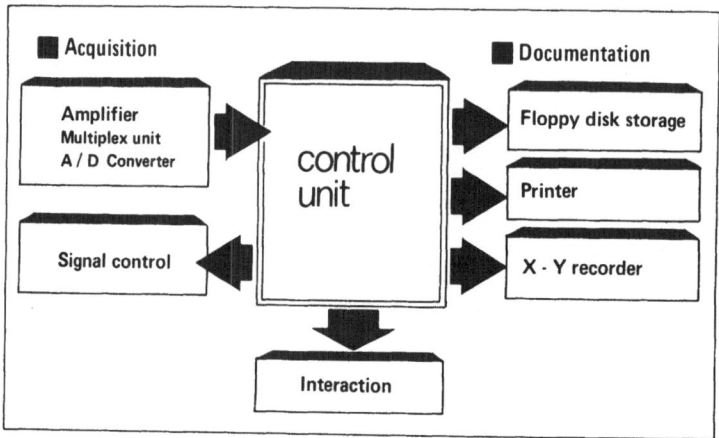

Fig. 1. Schematic diagram of the realized mapping device

SYSTEM FOR ACQUISITION AND PROCESSING OF MULTIPLE ECGS

Figure 1 shows a diagram representing the realized mapping
system with its main functional groups. The system based on a microprocessor has been designed for application of the measurement in the
coronary care unit. Much importance has been attached to the construction points that ensure the least possible strain for the patient and no disturbance of the therapeutic interventions. The
system must be easy to operate by the physicians as well as by the
nurses of the CCU. Therefore we reduced the number of control elements to a minimal necessary set of buttons that is placed on a
specially designed extensible panel. Ergonomic design principles
have been applied consequently.

Figure 2 shows more details of the realized system. The microprocessor supervises the whole system according to a special control
software. The main peripheral units work independently with a direct
memory access (DMA). The selected double density floppy disk mass
storage device offers large storage capacity and short access times
in order to ensure a sufficient throughput of digitized ECG-signals.

All of the thoracic electrodes are connected to a highly flexible plastic sheet. The transparent realization of the sheet allows
visual control of the tight contact between electrodes and the skin
surface. Because the measurement system was used first for automated
precordial ST-segment mapping, we chose a lead system similar to that
proposed by Maroko et al. (1972) with 48 thoracic electrodes and
standard leads I, II and III. All leads are sampled simultaneously
during one heart complex. Up to now we either used only a 32 x 24 cm
region of the surface potentials for further analysis or made records by placing the plate on 4 adjacent thoracic positions sequen-

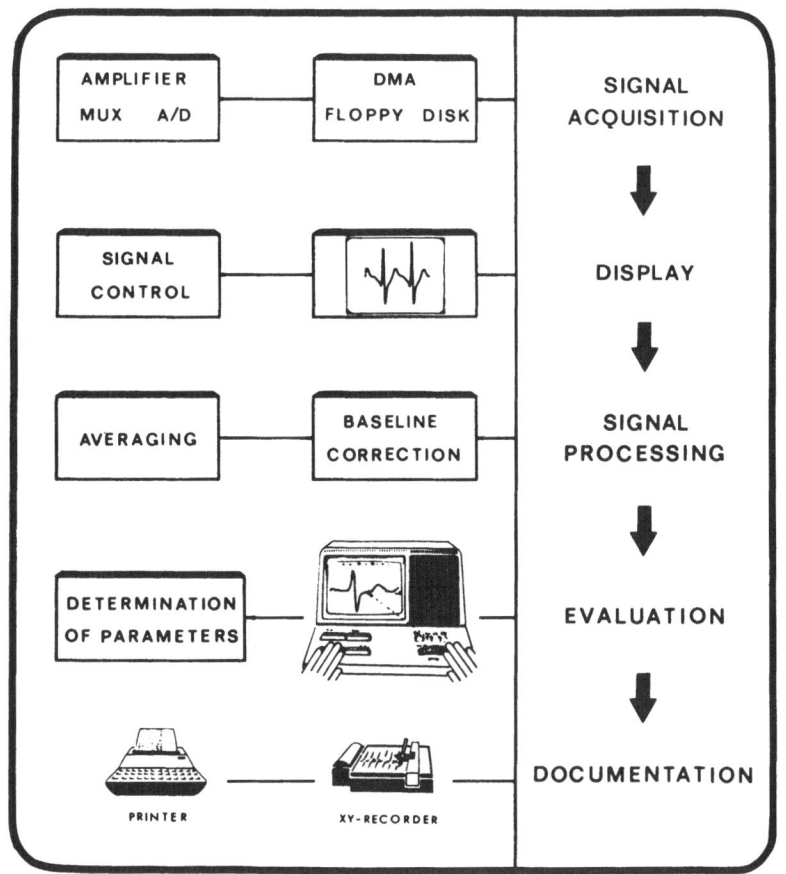

Fig. 2.   Diagram describing the different operation modes of the system

tially over several heart periods and then calculated the resulting total potential distribution by a time alignment of the different registrations. We do not intend to increase the number of electrodes over 5 - 6 but try to find the optimal locations. First investigations (Barr et al., 1971; Lux et al., 1976; Kornreich et al., (1978) have demonstrated, that for special diagnostical purposes there exist minimal lead electrode configurations that obtain the full information content by placing the electrodes in the optimal positions. The introduction of such new optimal lead systems has already been considered in the system design of the mapping device so that there will be no software conflicts when changing the electrode array.

We selected a sampling rate of 500 /sec per channel and 12 bit resolution for the analog to digital conversion. Each electrode is connected to an amplifier. The realized amplifier unit consists of

equal differential amplifiers with a voltage gain of 1000. All thoracic leads as well as standard leads I, II and III are amplified,
multiplexed, digitized and stored on the floppy disk. The specifica-
tions are described more explicitly in Hinsen et al., (1979). After
the sampling period, a quality control of the recorded ECG signals
can be performed by accelerated display of all signals on the CRT
screen. In an alternative mode the system automatically controls if
the signal quality is satisfying and marks the ECG intervals of lower
quality. Software routines for removing interferences of line frequency and baseline drifts are implemented. A representative heart
complex is derived from all beats recorded in the sampling period.
Relevant parameters e.g. ΣQ, ΣR, ΣST, NST, etc., are determined
automatically by using the pattern recognition methods or in borderline situations interactively by recalling the registered signals to
the CRT screen and marking all the relevant data points. For qualitative analysis records are plotted as isopotential contour maps or
as three-dimensional views of the surface potential distribution.
For quantitative analysis the system produces the measurement reports
and trend curves of all monitored parameters. All of the measurement
data and reports are stored on the diskette of the patient. They can
be printed or plotted immediately after a measurement or off line.

We used a separate data diskette for each patient. It obtains
additional personal information about the patient. This diskette has
to be initialized with the patient's data before the first acquisition and can be attached to his other clinical documents. All of the
status parameters of the system that are used for the formatted storage of the data are automatically updated, e.g. last acquisition
number, last report number, date, time, etc.

MONITORING OF PATIENTS WITH ACUTE MYOCARDIAL INFARCTION

During the last years the mortality in acute myocardial infarction could be reduced by an improved therapy of malignant arrhythmias and pump failures. The interest is now focused on minimizing
the final infarct size by therapeutical interventions. In order to
evaluate the different interventions, methods are required which
permit quantification of ischemia and necrosis. The method of the
body surface mapping and precordial ST-segment mapping seems to be
useful (Maroko et al., 1972; Reid et al., 1971).

Using our mapping system, we investigated the changes in the
surface potentials during 2 days after onset of the acute symptoms
in 12 patients with acute anterior myocardial infarction. Potential
maps were recorded at 2 hour intervals. Figure 3 shows a computer
plot of the registered signals after the averaging of a representative heart complex and elemination of baseline drifts. With the
method of precordial ST-segment mapping we investigated the course of
several calculated SUM-values and used this information for evalua-

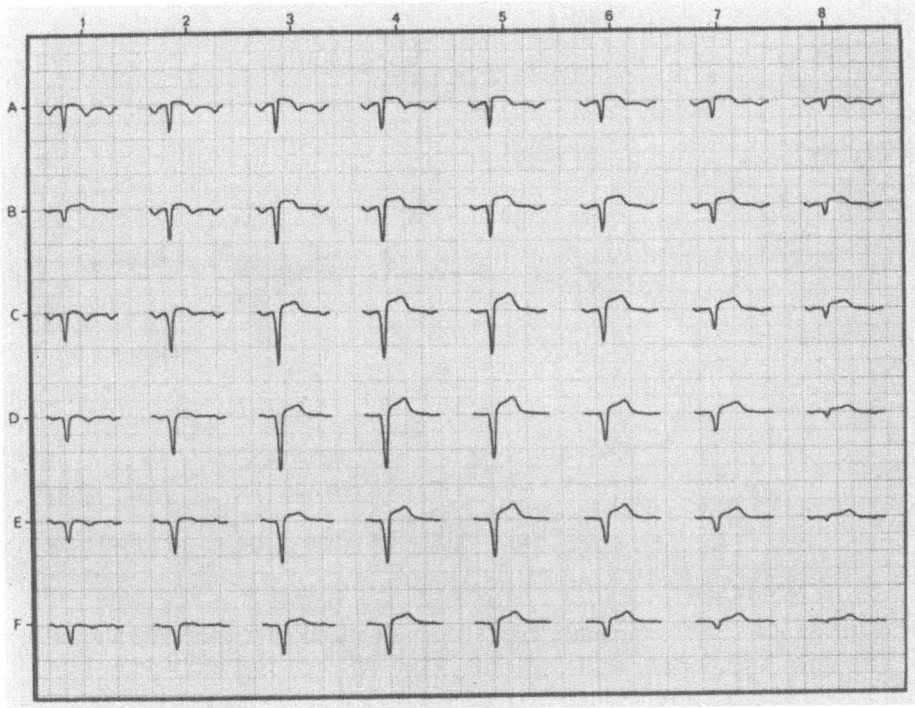

Fig. 3.   Plot of registered ECG signals. The local position of each
signal corresponds to the electrode array.

tion of the spontaneous course of myocardial infarction and for testing the effectiveness of therapeutic interventions. The trend of ST-
segment elevation was evaluated as the indicator of reversible myocardial ischemia, Q-wave development and R-wave reduction as indica-
tors of irreversible myocardial damage.

In Figure 4 the trend curves of sum values ΣQ, ΣR and ΣST are
shown as well as the MBCK serum curve. The course of ST-segment
elevation, Q-wave development and R-wave reduction indicates an uncomplicated course of acute anterior myocardial infarction.

The parameters are calculated in the following manner:

a) $\displaystyle\sum_{n=1}^{48} Q_n$ $\displaystyle\sum_{n=1}^{48} R_n$ $\displaystyle\sum_{n=1}^{48} ST_{60n}$

The sum of all ST-segment elevations 60 msec after the spike of
the S-wave and the sum of all Q-wave and R-wave amplitudes.

b)   The number and position of all leads having an ST-segment
above a defined amplitude ΔU

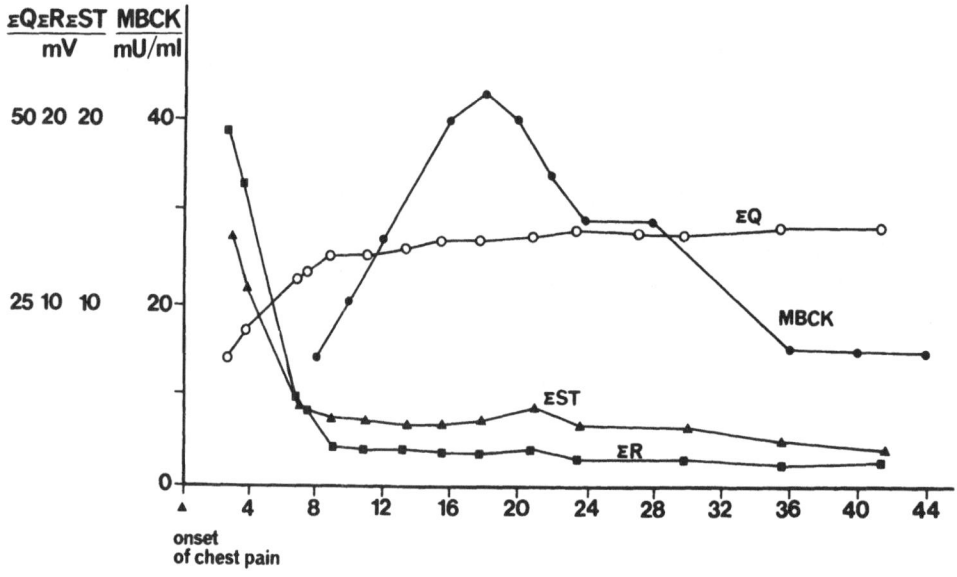

Fig. 4. Course of ST-segment elevation, Q-wave development and R-
         wave reduction in a patient with uncomplicated acute an-
         terior myocardial infarction.

c) $\sum\limits_{n=i}^{i+k} Q_n$ $\sum\limits_{n=i}^{i+k} R_n$ $\sum\limits_{n=i}^{i+k} ST_{60n}$

The sums of amplitudes in a region around the highest ST-segment
elevation.

All parameters are calculated automatically by the mapping
system.

There is an initial reduction of the ST-segment elevation during
the first 8 - 12 hours after onset of the chest pains. Within the
same period Q-waves appear and R-waves disappear. In uncomplicated
infarctions there are only small further changes during the first 2
days after onset of acute symptoms. In patients with new angina,
however, we could often see a new increase of ST-segment elevation
and a new development of Q-waves and further reduction of R-waves,
followed by a delayed increase of MBCK. These ECG changes are in-
dicators for an extension of the infarction.

Additional information about the myocardial injury could be obtained by the analysis of isopotential maps and three dimensional
views of body surface potentials (Fig. 5). These plots were produced automatically by the mapping system for each desired instant of

B.H./21·27/11-10-79

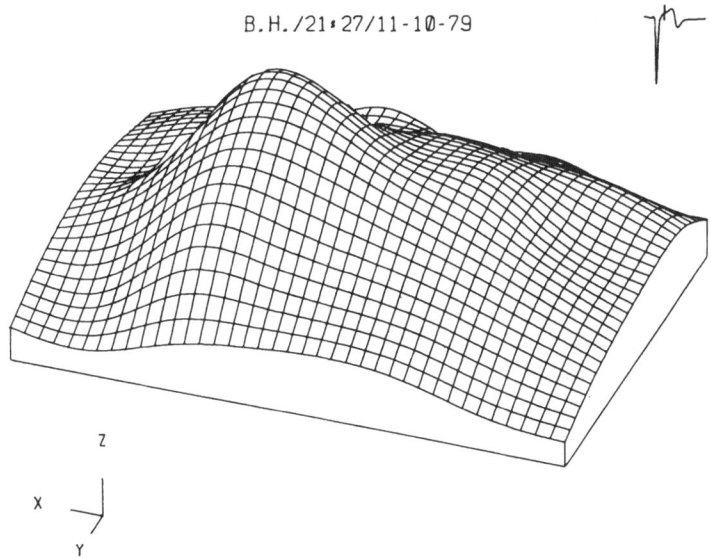

Fig. 5.   Three dimensional view of surface potentials 60 msec after
the end of QRS (patient with acute myocardial infarction).

the recorded heart complex. Using a sampling rate of 500/sec the
maximum time resolution for these maps is 2 msec. Every two hours,
together with precordial ST-mapping, we plotted a series of isopotential and 3D maps of the surface potentials during ventricular depolarization and repolarization. Maps were plotted at intervals of
2 msec duration during QRS and intervals of 10 msec during ST and T-
wave. Analysis was performed by visual observation of the areas of
positive and negative potentials and of the movement of these potential areas during the heart period.

AUTOMATION OF MAPPING DURING STRESS TESTS

Exercise electrocardiography is one of the most reliable non-
invasive clinical methods for the detection and quantification of
coronary heart diseases. However its sensitivity is not yet satisfying. In many patients with abnormal coronary arteries, documented
by selective coronary arteriography, the conventional exercise ECG
remains within normal limits. Several authors attempted to improve
the diagnostic value of the exercise ECG by means of computer processing (Simoons et al., 1977) or by recording of precordial exercise maps with multiple electrode arrays (Fox et al., 1979). We
combined both ideas by using our computerized mapping system. In a
comprehensive study accompanying the routine stress tests (conventional ECG) in our exercise ECG center we recorded and analysed
exercise maps from 135 consecutive patients with suspected coronary
heart diseases.

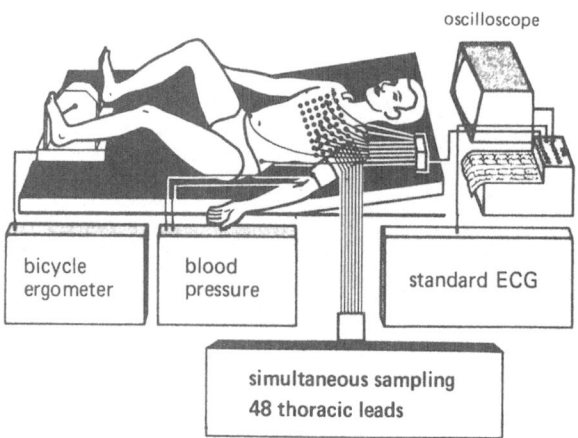

Fig. 6. The stress tests are performed in supine position using a
bicycle ergometer. Mapping is performed in addition to the
registration of standard ECG and blood pressure.

        Because of the simultaneous registration of all leads, the map-
ping system allows repetitive recordings at short intervals. The
implemented exercise ECG-software performs special tasks: noise reduction, baseline correction and extraction of a representative complex. In this way, artefacts are removed and the signal quality is
considerably improved, a main prerequisite for an exact quantitative
analysis.

        The exercise tests were performed on a bicycle ergometer in
supine position (Fig. 6). Workload was stepwise incremented. The
heart rate, blood pressure and standard ECG were monitored. The patients had been motivated to continue the test until the predicted
heartrate was reached or until fatigue, unless symptoms like angina
pectoris occurred. Standard ECG as well as surface maps were recorded during the resting condition and then at intervals of 1 minute
during the exercise until the highest workload was reached, immediately after the exercise as well as 1 and 6 minutes afterwards. The
signal acquisition software allows continuous ECG registration with
a sampling rate of 500/sec per channel. After a total sampling time
of 10 seconds the data diskette side must be changed. The diskettes
can be used on both sides, that means 20 sec of ECG data can be
stored on one diskette. After the registration, all heart complexes
of the sampling period were averaged. In each heart beat either the
T-P or the P-Q segment was used as isoelectric line. The choice de-
pended on which of these segments remained identifiable before and
after exercise. After the digital preprocessing of the sampled ECG
data we computed several parameters for further analysis. ST-segment
depressions were measured at 30, 60 and 80 msec after the end of QRS

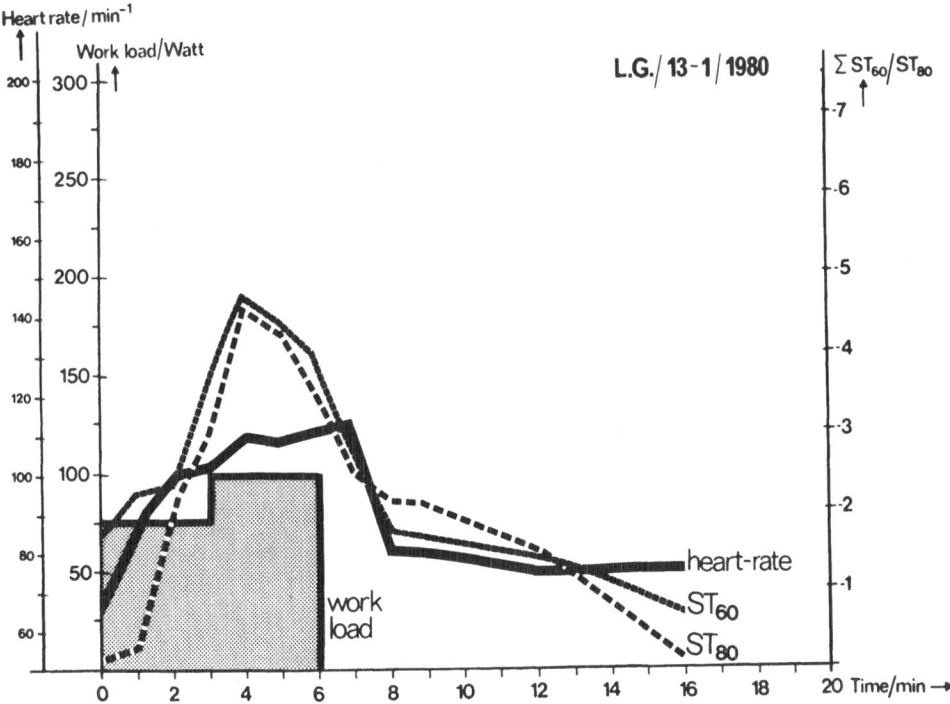

Fig. 7. Work load, heart rate and SUM ST60, SUM ST80 during a stress
test: patient with ischemic reaction.

(junction point). Both up- and down-going ST slopes and integrals
were then computed. Each integral was computed as the surface area
between baseline and negative part of the ST segment from the junction point to the X-point, the crossing of baseline and the ST-
segment. Immediately after the evaluation, the measurement report
is displayed on the CRT screen. It shows the patient's personal
data, the amplitudes, slopes and integral values of the momentaneous
measurement and the course of all parameters computed from previous
registrations.

Figure 7 shows the course of heartrate as well as of ST60 and
ST80 sum values. These sum values were computed by summation of the
ST-amplitudes of all leads showing an ST depression of at least 0.1
mV. The course of ΣST60 and ΣST80 in the figure 7 indicates a typical ischemic reaction during the exercise. The test was stopped
after 6 minutes because of severe chest pain and ST depression in the
control leads on the monitor.

In addition to the described parameters, the mapping system
produces isopotential surface maps and three dimensional views of

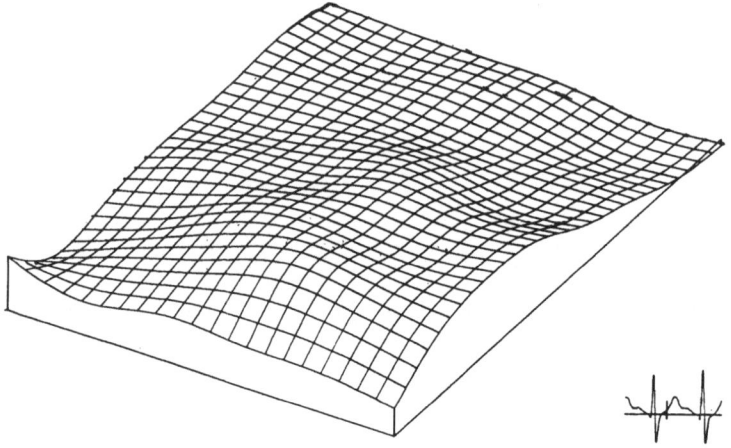

Fig. 8. Distribution of repolarization potentials (ST60) from a
         normal young male patient during exercise.

the repolarization potentials. Typical exercise induced patterns of
repolarization in the surface potentials are highly specific for
coronary heart disease (Figures 9 and 10). Figure 8 shows the precordial distribution of repolarization potentials 60 msec after the
end of QRS (ST60) during exercise for a young male patient. The potential is slightly positive all over the anterior chestwall. Figure
9 shows a distribution of precordial potentials recorded from a patient with angina pectoris during exercise. The depression of potentials on the left anterior chestwall is an indication of exercise in-

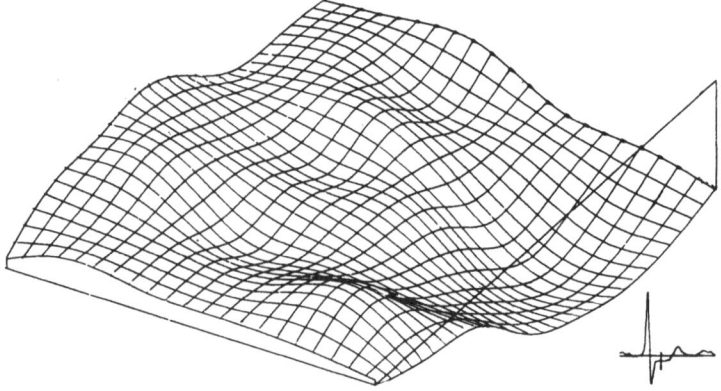

Fig. 9. Repolarization potentials (ST60) from a patient with
         ischemic reaction during exercise.

G.L./13-6/01-02-80

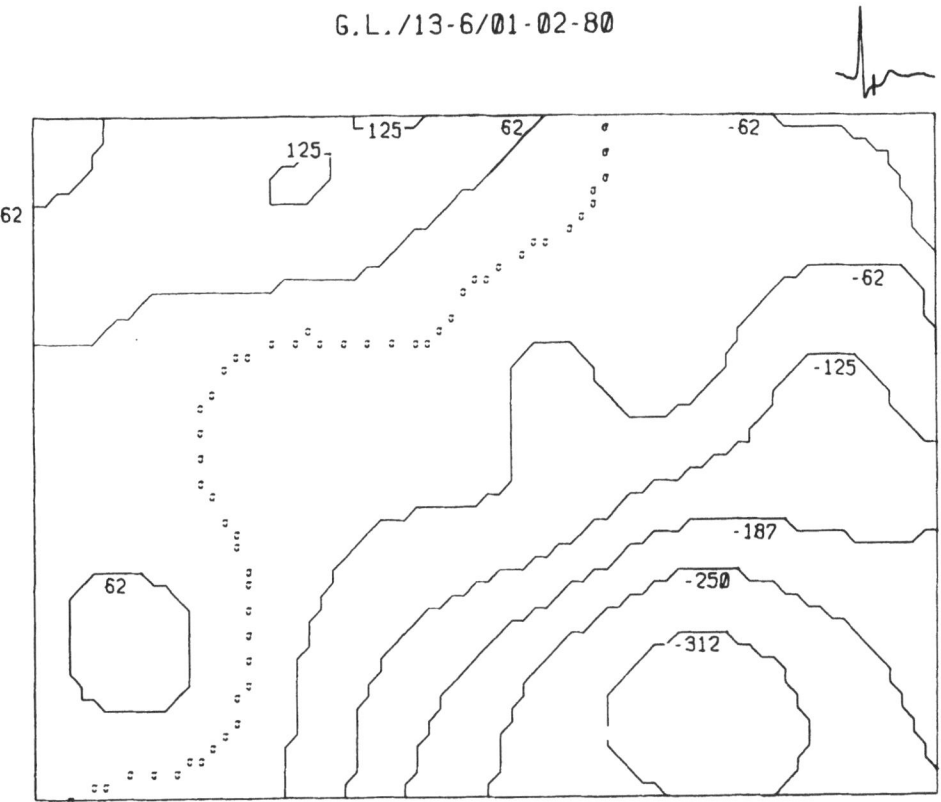

Fig. 10.   Isopotential map of the repolarization potentials shown in
           Figure 9.

duced myocardial ischemia. Figure 10 contains the same information
in another form: the potential distribution is presented as an iso¬
potential contour map. The precordial area of exercise induced patterns, the severity and the time course of electrocardiographic changes during the exercise give valuable additional information for the
diagnoses of coronary heart disease.

PROSPECTS

     The described mapping device allows comprehensive clinical
studies with a large number of patients. The reported applications
have documented that mapping provides information missed by conventional ECG techniques. In further studies we will investigate if
the mapping technique can be used to improve the diagnosis of other
pathologic cardiac states. These investigations will help us to find
new clinically relevant criteria and to optimize our electrode configurations as well as our processing procedures.

REFERENCES

Abildskov, J. A., Burgess, M. J., Urie, P., Lux, R. L., and Wyatt,
    R. F., 1977, The unidentified information content of the
    electrocardiogram, Circ. Res., 40:3-7.
de Ambroggi, L., Taccardi, B., Macchi, E., 1976, Body surface maps of
    heart potentials: Tentative localization of pre-excited
    areas in forty-two Wolff-Parkinson-White patients, Circulation
    54: 251-263.
Barr, R. C., Spach, M. S., and Herman-Giddens, G. S., 1971, Selection
    of the number and positions of measuring locations for elec-
    trocardiography, IEEE Trans. Biomed. Eng., 18: 125-138.
Cottini, C., Dotti, D., Gatti, E., and Taccardi, B., 1972, A 240-
    probe instrument for mapping cardiac potentials, in: "Proc.
    Satellite Symp. 25th Int. Congr. Physiol. Sci.," Presses
    Académiques Européennes, Brussels: 99.
Flowers, N. C., Horan, L. G., and Johnson, J. C., 1976, Anterior in-
    farctional changes occurring during mid and late ventricular
    activation detectable by surface mapping techniques, Circula-
    tion, 54: 906-913.
Fox, K. M., Selwyn, A. P., and Shillingford, J. P., 1979, Precordial
    electrocardiographic mapping after exercise in the diagnosis
    of coronary artery disease, Am J. Cardiol., 43:541-546.
Hinsen, R., von Essen, R., Silny, J., Merx, W., Rau, G., and Effert,
    S., 1979, Monitoring of myocardial ischemia and necrosis in
    acute myocardial infarction, in: "IEEE Proc. Computers in
    cardiology," IEEE Computer Society, Geneva.
Kornreich, F., 1973, The missing waveform information in the ortho-
    gonal electrocardiogram (Frank Leads). Part I: Where and how
    can missing waveform information be retrieved? Circulation,
    48:984-995
Kornreich, F., Smets, P., and Kornreich, J., 1978, About the unique-
    ness of 'optimal', 'total' or 'maximal' lead systems, Adv.
    Cardiol., 21:138-146.
Lux, R. L., Smith, C., Burgess, M. J. Wyatt, R. F., Vincent, G. M.,
    and Abildskov, J. A., 1976, Limited lead system for estima-
    ting total body surface potential maps of normals and patients
    with myocardial infarction, Am. J. Cardiol., 37:152.
Maroko, P. R., Libby, P., Covell, J. W., Sobel, B. E., Ross, J., and
    Braunwald, E., 1972, Precordial ST segment elevation mapping:
    an atraumatic method for assessing alterations in the extent
    of myocardial ischemic injury, Am. J. Cardiol., 29: 223-230.
Reid, D. S., Pelides, L. J., and Shillingford, J. P., 1971, Surface
    mapping of RST segment in acute myocardial infarction, Br.
    Heart J., 33: 370-374.
Simoons, M. L., Ascoop, C. A., Block, P., Distelbring, C. A., de
    Lang, P. A., and Vinke, R., 1977, Computer processing of exer-
    cise electrocardiograms: A cooperative study, in: "Trends in
    computer-processed electrocardiograms," North Holland Publi-
    shing Company, Amsterdam, pp. 383-384.

Taccardi, B., 1963, Distribution of heart potential on the thoracic surface of normal human subjects, Circ. Res., 12: 341-352.

Tiberghien, J., Steenhaut, O., Lenders, P., Block, P., Kornreich, F., Reygaert, P., and Crabbe, J. P., 1976, An 128-Channel electrocardiographic recorder, Adv. Cardiol., 16: 128-132.

Vincent, G. M., Abildskov, J. A., Burgess, M. J., Millar, K., Lux, R. L., and Wyatt, R. F., 1977, Diagnosis of old inferior myocardial infarction by body surface isopotential mapping, Am. J. Cardiol., 39: 510-515.

Wyatt, R. F., and Lux, R. L., 1974, Application of multiplexing techniques in the collection of body surface maps from single complexes, Adv. Cardiol., 10: 26-32.

Yamada, K., Toyama, J., Wada, M., and Sugiyama, S., 1975, Body surface isopotential mapping in WPW syndrome: Noninvasive method to determine the localization of the accessory atrioventricular pathway, Am. Heart J., 90:721-734.

SURFACE MAPPING CHARACTERISTICS OF LEFT FASCICULAR BLOCKS

I. Préda, Gy. Kozmann* and Z. Antalóczy

Postgraduate Medical School
2nd Medical Clinic, Budapest, Hungary
*Central Research Institute for Physics
Budapest, Hungary

INTRODUCTION

When Rosenbaum et al. (1968) decided upon the criteria for the diagnosis of left anterior hemiblocks, he remarked that it was difficult to draw a precise dividing line between ordinary left axis deviation and left axis deviation due to left anterior hemiblocks. The existence of different degrees of left anterior hemiblocks has been recognized from the outset: different left anterior hemiblock patterns may occur, for example in the same patient, during aberrant ventricular conduction of premature supraventricular beats induced experimentally by programmed atrial stimulation (Kulbertus et al., 1976). As suggested by Durrer et al. (1966) the forme fruste endocardial cushion defect can produce left axis deviation probably due to abnormal development of the conducting system. The left posterior hemiblock as well as the left septal fascicular block do not behave as clearcut phenomena, and the different authors agree on their deceptive and non-specific nature (Rosenbaum et al., 1972); however in the majority of cases the histological findings are consistent with the known electrocardiographic criteria (Demoulin and Kulbertus, 1972).

The different electrocardiographic patterns of the left sided conduction system can be explained by its great variability as was demonstrated by Demoulin and Kulbertus (1972), but in spite of this it seems also to be desirable to characterize more precisely the different patterns and possibly to distinguish the normal variables from the pathological ones.

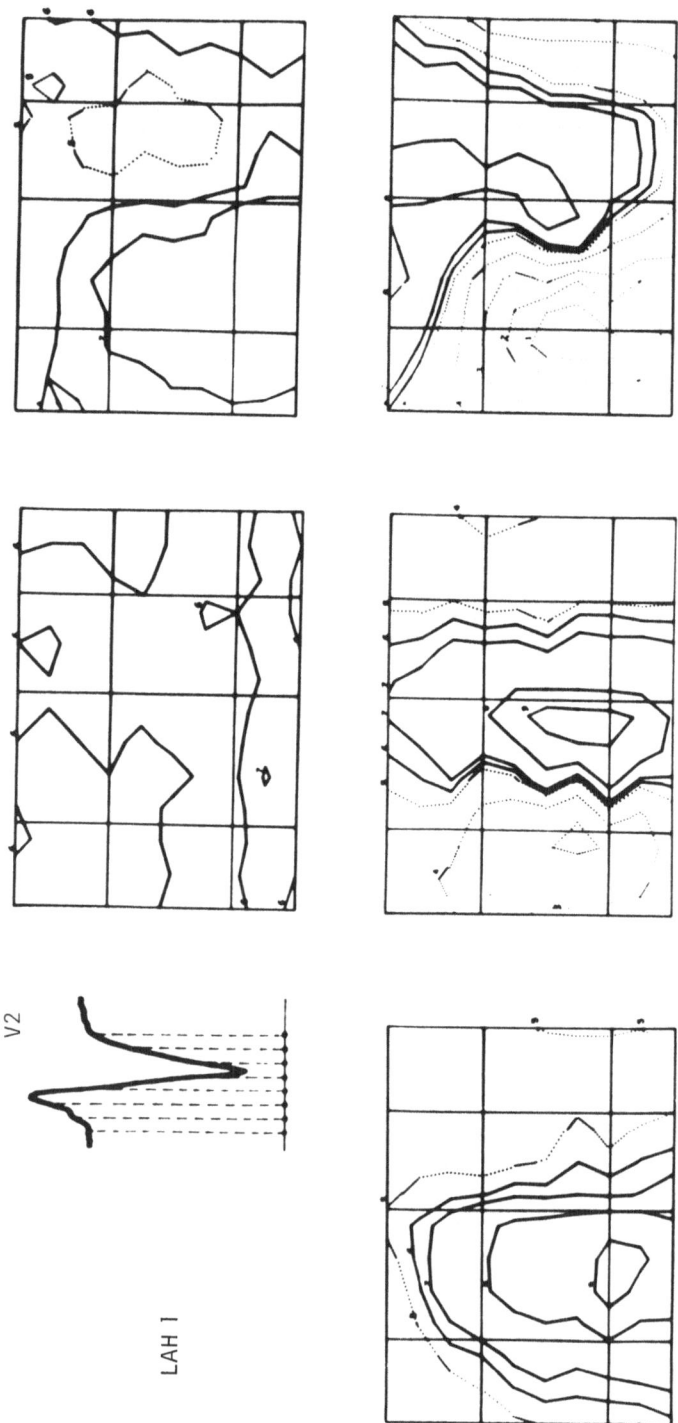

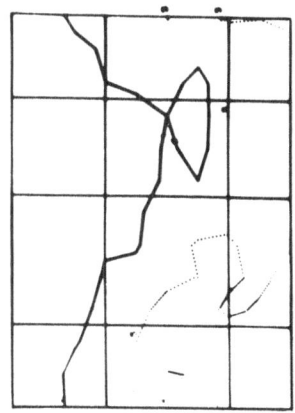
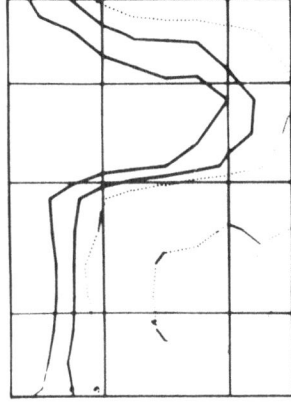
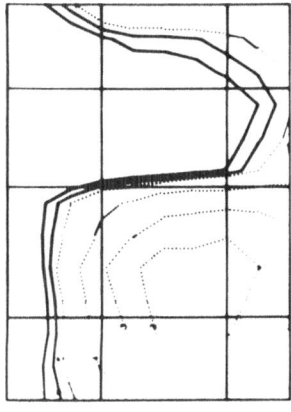

Fig. 1. Fig. 1 illustrates a representative time course of Type 1 LAH maps at 8 equidistant intervals during the QRS complex. The left side of the map illustrates the anterior, the right side the posterior chest surface. The dotted lines represent negative values, the solid line No. 6 stands for the zero potential, and the higher values of solid lines are representing positive equipotential contours. (For explanation see text.)

METHODS

The present study deals with the surface mapping characteristics of 25 subjects revealing the classical electrocardiographic patterns of one of the left sided fascicular blocks. In all of the patients previous myocardial infarction, congenital valvular disease, pulmonary emphysema from the patient's history and clinical point of view (ECG, phonocardiography, mechanocardiography, echocardiography, laboratory examinations) have been excluded. The surface mapping measurements were derived from 120 unipolar chest electrodes (Kozmann et al., 1976; Préda et al., 1979) and evaluated in the form of isopotential surface maps for every 2 msec of a heart cycle by a TPA/i computer. To assess the possible alterations in ventricular activation, the surface maps were related to normal subjects, and the classical electrocardiological criteria (ECG, VCG and polar-coordinates) were also taken into account (Antalóczy, 1972; de Padua et al., 1978; Hoffman et al., 1976; Nakaya et al., 1978). For an explanation of the different maps during the cardiac depolarization the known data of the human cardiac activation in normal and pathological cases (van Dam, 1976; Durrer et al., 1970; Medrano et al., 1970) were analysed and the migration of the main maximum (Taccardi et al., 1976) and the construction of average differential maps (McLaughlin et al., 1974; Toyoshima et al., 1974; Kozmann and Préda, in this volume) were used as a feature extraction.

RESULTS

A) Left Anterior Hemiblock

22 patients exhibited the electrocardiographic criteria of left anterior hemiblock (QRS duration less than 120 msec; QRS axis in the frontal plane -30°, or more negative). Based on a careful inspection of the maps and the migration of the main maximum 4 more or less characteristic subtypes can be distinguished:

Type 1. left anterior hemiblock (LAH) maps were represented by 5 subjects (Fig. 1). At the beginning of the QRS complex the first maximum was observed in the lower part of the middle of the anterior chest wall, whereas the first significant minimum appeared on the middle and upper third of the back, and during the building up of the maximum it migrates towards the right axillary line. In the second third of the depolarization the anterior maximum gradually migrates towards the left side of the anterior chest wall, and increases in area towards the left shoulder and the left upper part of the chest, while the minimum approaches the right side and the middle of the anterior thoracic wall. In the terminal third of the QRS complex the maximum, or separated maxima were characteristically observed in the middle and upper part of the posterior chest wall, the principal maximum migrates upwards, and turning around the right shoulder it terminates in the middle of the upper part of the anterior thoracic surface.

The migration of the main maximum of each of the 5 subjects belonging to the Type 1 LAH maps is depicted on the Fig. 2/a. As it is shown, the migration of principal maxima in all of the cases is considerably homogeneous, the mean QRS duration is normal, their mean age was relatively low.

Type 2 LAH maps were also represented by 5 subjects, 4 of them exhibited clinical signs of coronary heart disease. One of the main differences is (Fig. 2b), that the first maximum occurs in these patients in the middle third of the sternal region; just as it is habitual in normal subjects (Taccardi, 1963). In the second third of the QRS complex, in main features, the migration of the maxima and minima is similar to the Type 1 LAH surface maps, and the terminal maximum also appears on the higher part of the anterior chest wall.

As it is depicted on the Fig. 2c the main maximum in the Type 3 LAH cases (7 patients, all of them suffered from coronary heart disease or congestive cardiomyopathy) migrates inhomogeneously; the maximum initiates from the lower part of the anterior chest wall and travelling by great oscillations it terminates on the posterior chest wall but rather irregularly from person to person.

Type 4 LAH was represented by 3 cardiac patients (Fig. 2d). The first maximum of these maps appeared on the upper or on the middle part of the sternal region. The second and third part of the QRS complex are in their main characteristics similar to the Type 3 maps, but the multiple extrema of different location on the maps are perhaps distinguishing features from the other types of LAH maps.

B) Left Septal Fascicular Block

One patient was observed to fulfil the ECG criteria (de Padua et al., 1978; Nakaya et al., 1978) of left septal fascicular block (LSF). It can be seen (Fig. 3) that the first maximum appears on the left upper side of the midsternal border, gradually augmenting in voltage and area opposite to a minimum of the posterior chest wall. In the second third of the QRS complex a new minimum appears on the right upper part of the thorax and the maximum divides into two positive peaks and slowly shifts downwards and to the left. In the terminal phase of depolarization the maxima migrate laterally and backwards, - opposite to minimum or minima on the anterior chest wall.

C) Left Posterior Hemiblock

Two patients exhibited the ECG criteria (Rosenbaum et al., 1968; 1972) of left posterior hemiblock (LPH). In these cases (Fig. 4) the

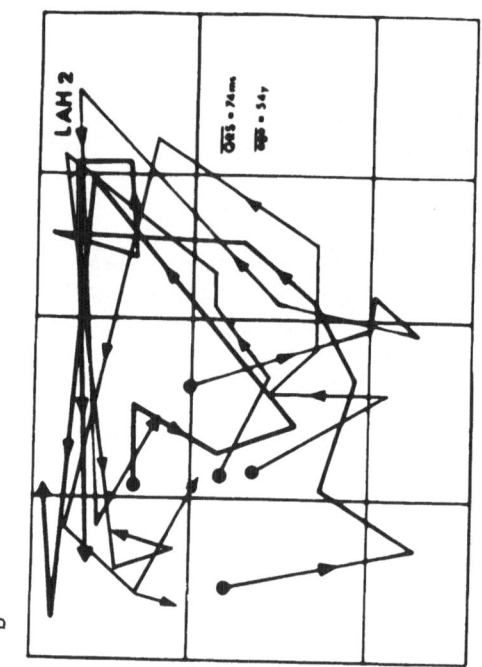

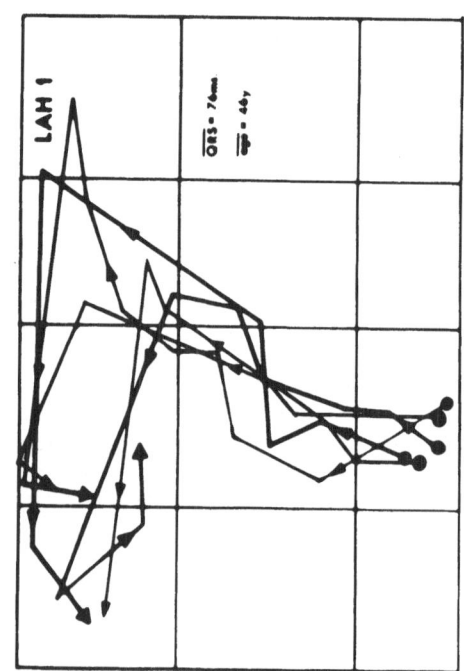

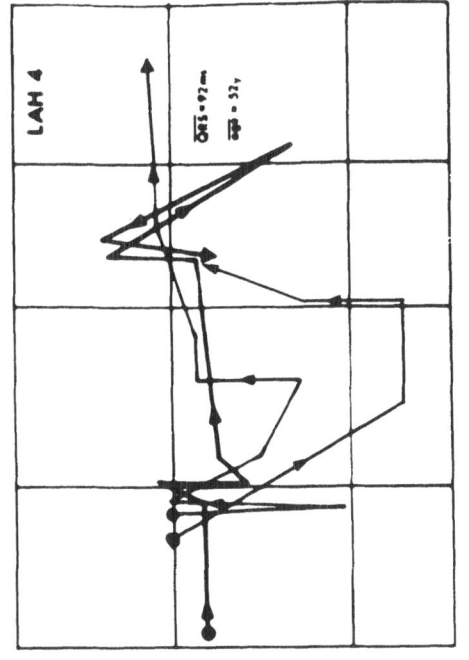

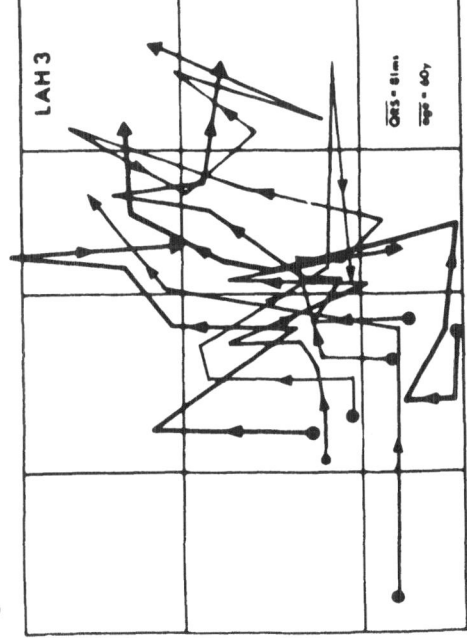

Fig. 2. Schematic representation of the migration of the main maxima during the QRS complex in different types of LAH. The left side of the figures illustrates the anterior, the right side the posterior chest surface.

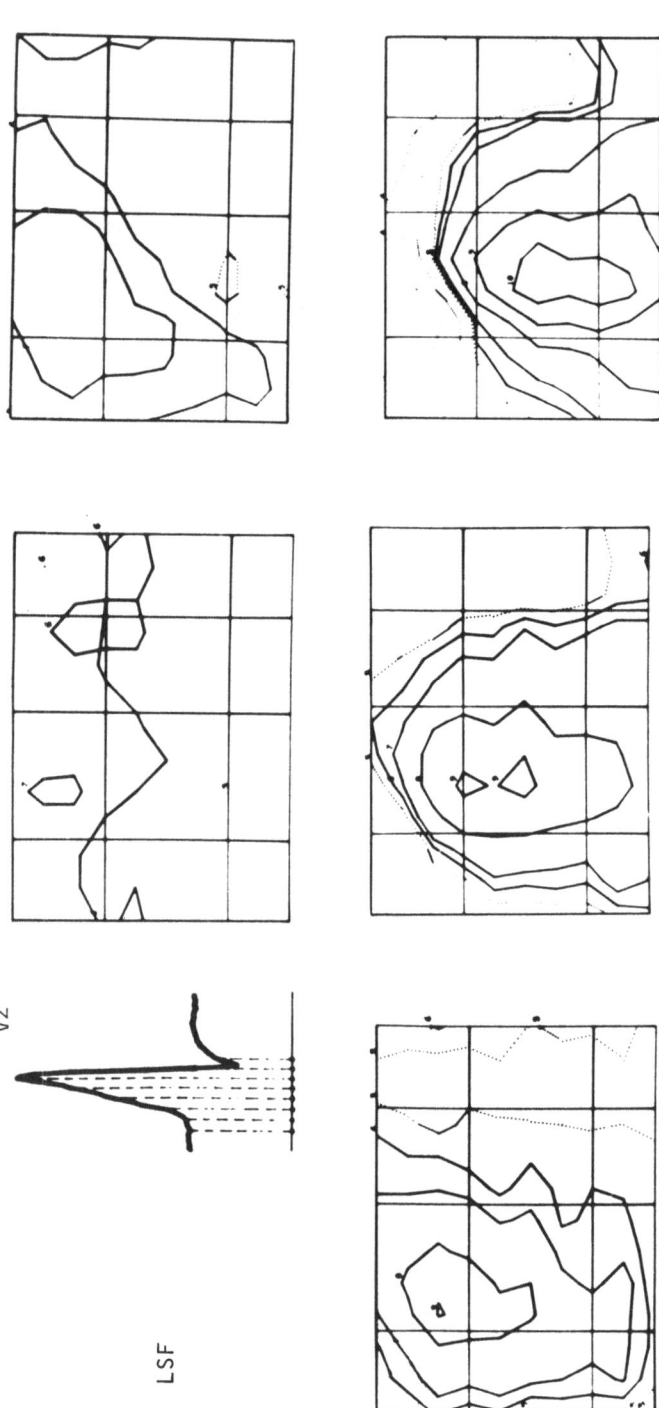

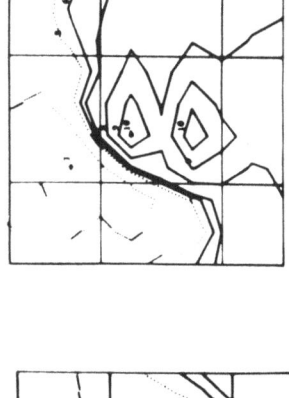

Fig. 3.   Representative time course of the LSF map. The left side of the maps illustrates the anterior, the right side the posterior chest surface. The dotted line represents negative values, the solid line No. 6 stands for the zero potential, and the higher values of solid lines are representing positive equipotential contours. (For explanation see text.)

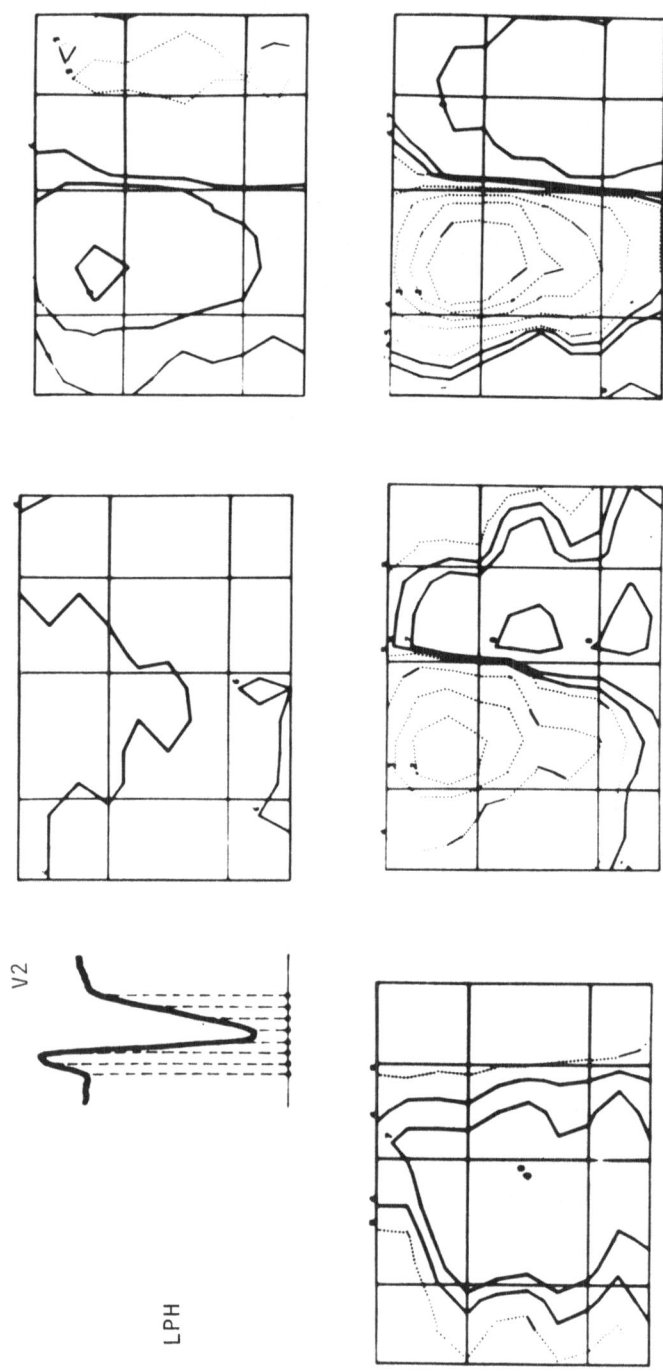

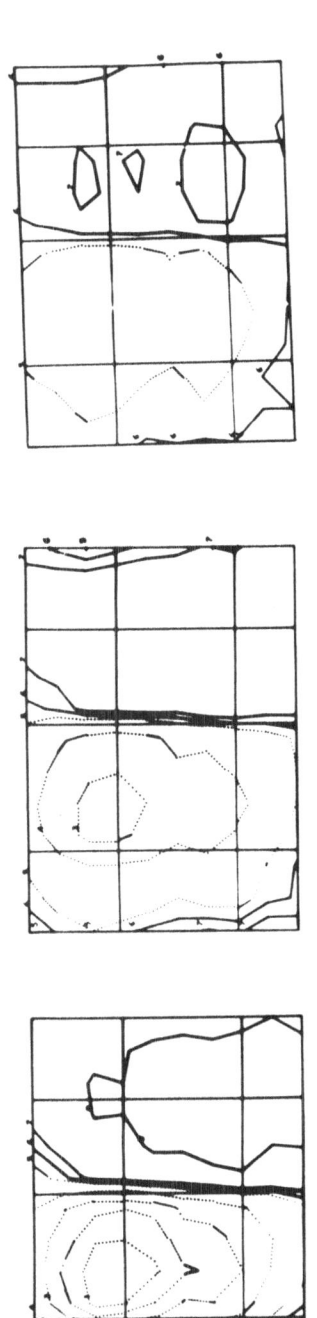

Fig. 4.   Representative time course of LPH maps. The left side of the maps illustrates the anterior, the right side the posterior chest surface. The dotted line represents negative values, the solid line No. 6 stands for the zero potential, and the higher values of solid lines are representing positive equipotential contours. (For explanation see text.)

first maximum initiates normally on the left side of the upper midsternal border opposite to a minimum on the back. At the beginning
of the second third of the QRS, the anterior maximum in both cases
was distributed and migrated downwards and to the left, whereas a
growing minimum appeared on the right upper side of the anterior
chest wall probably reflecting a still depolarized area. In the
middle of the depolarization the distributed maximum migrates gradually to the left and posteriorly, while in the terminal phase of
depolarization the multiple maxima on the back may be related to the
terminal activation wavefronts depolarizing the different parts of
the posterior epicardial surface.

CONCLUSIONS

The results of the study suggest, that the surface mapping description and interpretation of left sided fascicular blocks provides
a new and detailed information for their further classification.
Simple methods of feature extraction like the representation of the
main maximum's migration and difference maps are useful tools for the
interpretation of the disturbed activation process. LAH-s may be
differentiated into 4 subtypes; Type 1 LAH maps are probably representing healthy subjects with a form fruste abnormal development of
a left sided conducting system. Due to different evidences, Type 3
and Type 4 LAH represent unambiguously diseased intraventricular conduction.

REFERENCES

Antalóczy, Z., 1972, Investigation on the electrical activity of the
        heart, Medicina, Budapest (in Hungarian).
van Dam, R. Th., 1976, Ventricular activation in human and canine
        bundle branch block, in: "Conduction System of the Heart,"
        H. J. J. Wellens, K. I. Lie, and M. J. Janse, eds., Leiden,
        pp. 377-384.
Demoulin, J. C., and Kulbertus, H. E., 1972, Histopathological ex-
        amination of concept of left hemiblock, Brit. Heart J., 34:
        807-814.
Durrer, D., Roos, J., and van Dam, R. Th., 1966, The electrogenesis
        of the electrocardiogram of patients with ostium defects
        (ventral atrial septal defects), Amer. Heart J., 71:642-650.
Durrer, D., van Dam, R. Th., Freud, G. E., Janse, M. J., Meijler,
        F. L., and Arzbaecher, R. C., 1970, Total excitation of the
        isolated human heart, Circulat. Res., 41: 899-913.
Hoffman, I., Mehte, J., Hilsenrath, J., and Hamby, R. L., 1976,
        Anterior conduction delay: a possible cause for prominent
        anterior QRS forces, J. Electrocardiol., 9:15-21.
Kulbertus, H. E., de Levalrutten, Fr., and Casters, P., 1976, Vec-
        torcardiographic study of aberrant conduction, anterior dis-

placement of QRS; another form of intraventricular block, Brit. Heart J., 38: 549-557.

Kozmann, Gy., Préda, I., Shakin, V. V., Szlávik, F., and Antalóczy, Z., 1976, Computer-aided method for the comparison of surface potential and acceleration maps, in: "Computers in Cardiology," Washington University, St. Luis, pp. 29-35.

Kozmann, Gy., and Préda, I., 1981, Estimation of cardiac excitation on the basis of stimulus response functions and epicardial activation isochrones, in this volume.

McLaughlin, V. W., Flowers, N. C., Horan, L. G., and Killam, H. A. W. 1974, Surface potential contribution from discrete elements of ventricular wall. A closed chest, postmortem-documented prospective study, Amer. J. Cardiol., 34: 302-308.

Medrano, G. A., Brenes, C. P., De Micheli, A., Cisneros, F., and Sodi-Pallares, D., 1970, The anterior subdivision block of the left bundle branch of His. 1. The ventricular activation process, J. Electrocardiol., 3: 7-11.

Nakaya, Y., Hiasa, Y., Murayama, Y., Ueda, S., Nagao, T., Niki, T., Mori, H., and Takashima, Y., 1978, Prominent anterior QRS force as a manifestation of left septal fascicular block, J. Electrocardiol., 11:39-46.

de Padua, F., des Reis, D. D., Lopes, V. M., Pereira Miguel, I., Lopes, M. G., da Silva, T. M., and de Padua, I. P., 1978, Left median hemiblock - a chimera? Adv. Cardiol., 21:242-248.

Préda, I., Bukosza, I., Kozmann, Gy., Shakin, V. V., Székely, A., and Antalóczy, Z., 1979, Surface potential distribution on the human thoracic surface in left bundle branch block, Japanese Heart J., 20:7-21.

Rosenbaum, M. B., Elizari, M. V., and Lazzari, J. O., 1968, "Los Hemibloqueos," Paidos, Buenos Aires, Argentinia.

Rosenbaum, M. B., Elizari, M., Lazzari, J. O., Nau, G. I., Halpern, M. S., and Levi, R. J., 1972, The differential electrocardiographic manifestations of hemiblocks, bilateral bundle branch blocks and trifascicular blocks, in: "Advances of Electrocardiography," Shlant, R., Hurst, J. W., eds., Grune and Stratton, New York, p. 145.

Taccardi, B., 1963, Distribution of heart potentials on the thoracic surface of normal human subjects, Circulat. Res., 12:341-350.

Taccardi, B., de Ambroggi, L., and Viganotti, C., 1976, Body-surface mapping of heart potentials, in: "Theoretical Basis of Electrocardiology," C. V. Nelson and D. B. Geselowitz, eds., Clanderon Press, Oxford, pp. 436-466.

Toyoshima, H., Sugiyama, S., Wada, M., Sugenoya, J., Toyama, J., and Yamada, K., 1974, Sequential change in the difference of potential distribution between a normal subject and simulated torso model, Japanese Heart J., 15:560-578.

SPATIAL VELOCITY IN THE SECOND PART OF QRS

IN THE ECG OF RABBITS

Radoslav Lolov

Department of Pathophysiology
Scientific Medico-Biological Institute
Medical Academy
Sofia, Bulgaria

The spatial velocity of the second part of QRS is a relatively new parameter which enriches the diagnostic capacity of the electro-cardiogram (Chorão de Aguiar et al., 1979). This parameter was de-fined at first by Hellerstein and Hamlin (1960) after experiments performed on dogs, and it was studied later on for healthy subjects (Draper et al., 1964; Chorão de Aguiar et al., 1979), as well as for patients with different cardiopathies (Gamboa et al., 1967; Chorão de Aguiar et al., 1976).

One of the possibilities for a future development of electro-cardiology consists in the improved interpretation of the bioelec-trical cardiac curves (Schaefer, 1976) with a view to enrich the information about the state of the heart. Particularly useful in this respect are experiments on animals, which permit modelling of the majority of cardiac damages observed in the clinical practice.

The aim of the present study is to determine the normal values of the spatial velocity (SV) of several successive time intervals in the second part of QRS in the Ecg of rabbits.

MATERIAL AND METHODS

We studied the electrocardiograms of 21 intact rabbits (10 female and 11 male), Belgian Giant breed, with an average body weight of 3500 g, which were anaesthetized with urethane (1.2 g/kg i.m.) and fixed in supine position with the forelegs placed behind the body. We used needle electrodes and a six-channel electrocardiograph RFT 6 NEK 401, at a recording speed of 200 mm/s, and an amplification of

10 mm/mV. The three scalar leads were recorded by a Frank's system. The recordings were processed with the method described by Chorão de Aguiar et al. (1979). The processing consists of determination of the end of the QRS-complex, plotting of the basic line by passing a straight line between two consecutive PR-intervals, measurement of the amplitude of the vectors (in mV) at every 2.5 ms, starting from the 10 ms vector and reaching the 22.5 ms vector before the end of the QRS-complex.

The spatial velocity was estimated by the equation of Hellerstein and Hamlin (1960):

$$SV = \sqrt{(Xa - Xb)^2 + (Ya - Yb)^2 + (Za - Zb)^2},$$

where a and b are the amplitudes in mV of the vectors at 12.5 and 10 ms, the vectors at 15 and 12.5 ms, etc., until reaching the vectors at 22.5 and 20 ms before the end of the QRS-complex. The values obtained were divided by 2.5 in order to obtain the measurement unit mV/ms. Each pair of vectors limited five time intervals each, which were designated by letters from A to E.

RESULTS AND DISCUSSION

The mean values of the results obtained and their errors are presented in Figure 1. The five time intervals studied for determining the spatial velocity of the different vectors of ventricular depolarization are situated between the vectors at 10 and 22.5 ms before the end of the QRS-complex. The duration of the entire QRS-complex in the different orthogonal recordings was found to vary between 35 and 45 ms for rabbits. In fact, the last interval we measured between the 20 and 22.5 ms vectors (E) is bounded with the peak of the R-wave. During this time interval SV reaches the highest values – 0.127 ± 0.015 mV/ms on an average, and here we have the greatest dispersion of the results from the different cases. With the advancing of the depolarization process, SV decreases. The graph plotted for SV reveals two sharp decreases (between the intervals B and A, and between D and C), which are statistically significant and designated by an asterisk. The SV value is the lowest (0.045 ± 0.005 mV/ms) in the interval A, i.e. between the vectors at 12.5 and 10 ms, where the dispersion of the results from the different cases is the smallest.

As it has been pointed out above, the parameter discussed has been characterized in the literature for healthy subjects and for patients with some cardiac diseases. In the experiments on animals this parameter has been studied only on dogs. Exact comparison between completely analogous moments of the ventricular depolarization for men and for the experimental animals would not be possible with

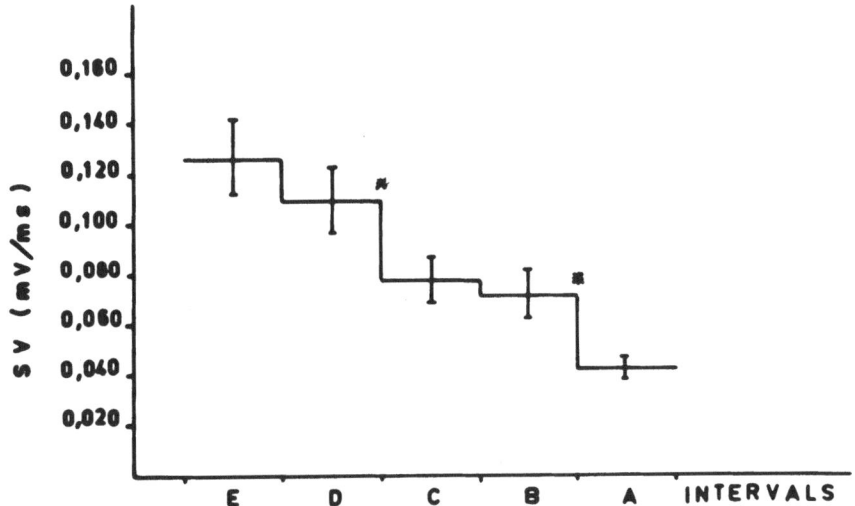

Fig. 1. Spatial velocity of the second part of QRS in rabbits.

the necessary precision. Therefore, we shall only point out that we established slightly higher SV values in the second part of the QRS-complex in rabbits. The resulting curve is steeper though analogous in character to the one described by the authors cited at the beginning of this article. We take into account the fact that the value of SV depends largely on the propagation velocity of the excitation process in the myocardium (Chorão de Aguiar et al., 1979). This gives us reasons to assume the probability that the relatively high SV values which we established in the initial intervals result from the relatively faster intraventricular conduction in rabbits.

Observations on SV have not yet elucidated sufficiently the diagnostic value of this parameter. More studies in this respect are needed for different kinds of cardiac diseases.

REFERENCES

Chorão de Aguiar, A. J., Moraid, M. E., Matos, N., and Guimaraes, H., 1976, Left axis deviation. Correlation between the orientation of the initial forces and the spatial velocity between the 27.5 and 30 msec vectors before the end of the QRS. A Cooperative study, Adv. Cardiol., 16:501-503.
Chorão de Aguiar, A. J., Guimaraes, H., and Raposo, A., 1979, The normal velocity in the second part of QRS (A cooperative study of 229 healthy individuals), J. Electrocardiol., 12(4):381-386
Draper, H. W., Peffer, C. J., Stallmann, F. W., Littmann, D., and Pipberger, H. V., 1964, The corrected orthogonal electro-cardiogram and vectorcardiogram in 510 normal men (Frank lead system), Circulation, 30:863-867.

Gamboa, R., Gupta, D. N., and White, N., 1967, Right bundle-branch
     block and the velocity of the electrocardiogram, Arch. Intern.
     Med., 120:286-290.
Hellerstein, H. K., and Hamlin, R., 1960, QRS component of the spa-
     tial vectorcardiogram and of the spatial magnitude and velo-
     city electrocardiograms of the normal dogs, Am. J. Cardiol.,
     6:1049-1056.
Schaefer, H., 1976, Possibilities of electrocardiography in the
     future, Adv. Cardiol., 16:18-26.

CONTRIBUTORS

Prof. Dr. H. Abel,
  6200 Wiesbaden,
  St. Joseph-Hospital,
  Solmsstr. 15 / FRG.

Prof. Dr. Z. Antalóczy,
  H-1389 Budapest XIII,
  Szabolcs u. 35,
  2nd Medical Clinic,
  Postgraduate Medical School /
  Hungary

Dr. G. Arisi,
  Parma,
  Via Gramzci 14,
  Ist. Physiol. Gen.,
  Univ. di Parma / Italy.

Dr. P. Barone,
  Roma
  Viale del Policlinico 137,
  Ist. Applic. del Calcolo,
  CNR / Italy.

Dr. S. Baruffi
  Parma
  Via Gramsci 14,
  Ist. Physiol. Gen.,
  Univ. di Parma / Italy.

Dr. O. Baum,
  142291 Pushchino,
  Moscow region,
  Inst. Biophys.,
  Akad. Sci. USSR / USSR.

Dr. V. Bonatti,
  Parma,
  Ospedale Maggiore,
  Divisiona di Cardiologia/Italy.

Dr. G. Botti,
  Parma,
  Ospedale Maggiore,
  Divisiona di Cardiologia /
  Italy

Dr. P. Chlebus,
  884 23 Bratislava,
  Sienkieviczova 1,
  Inst. Physiol. SAV / CSSR.

Dr. P. Ciarlini,
  Roma
  Viale del Policlinico 137,
  Ist. Applic. del Calcolo,
  CNR / Italy.

Dr. P. Colli Franzone,
  27200 Pavia,
  Corso Carlo Alberto 5
  Lab. di analis. num.
  CNR /Italy.

Dr. Z. Cserjes,
  H-1389 Budapest XIII,
  Szabolcs u. 35,
  2nd Medical Clinic,
  Postgraduate Medical School /
  Hungary.

Prof. Dr. R. van Dam,
  6500 HB Nijmegen
  Geert Grooteplein Zuid
  Sint Radboudziekenhuis
  Nederland

Dr. Z. Drška,
  180 85 Praha 8, Bulovka, Pav.11,
  Inst. Physiol. Regulat.
  CSAV / CSSR.

Prof. Dr. S. Effert,
  Aachen,
  Dept. Int. Med. I.
  RWTH Aachen,
  Goethestr. 27 / FRG.

Dr. Th. Eifrig,
  7010 Leipzig,
  Liebigstr. 27,
  Chair of Pathophysiology,
  Bereich Medizin,
  Karl-Marx-University / GDR.

Dr. M. Engst,
  1040 Berlin,
  Hessische Str. 3/4,
  Physiol. Inst. Humboldt-Univ./
  GDR.

Dr. L. Guerri,
  27100 Pavia,
  Corso Carlo Alberto 5,
  Lab. di analis. num. CNR /
  Italy.

Dr. R. Hinsen,
  Aachen,
  Goethestr. 27,
  Helmholtz-Inst. Biomed. Eng.,
  RWTH Aachen /FRG.

Dr. N. Kerin,
  Sinai Hospital of Detroit,
  6767 West Outer Drive
  Detroit MI 48235 / USA.

Dr. P. Kneppo,
  885 27 Bratislava
  Dúbravská cesta,
  Ustav merania, SAV / CSSR.

Dr. R. Lolov,
  1431 Sofia,
  1 G. Sofiiski,
  Med. Akad.,
  Dept. Pathophys. / Bulgaria.

Prof. Dr. E. Macchi,
  Parma,
  Via Universitá 12,
  Ist. di Matematica,
  Univ. de Parma / Italy.

Prof. Dr.P. Macfarlane,
  Glasgow G4 0SF,
  84 Castle Street,
  Glasgow Royal Infirmary,
  Scotland.

Dr. M. Maco,
  Bratislava,
  Dúbravská cesta,
  Inst. Measurement,
  Slov. Akad. Sci. / CSSR.

Dr. W. Merx,
  Aachen,
  Dept. Int. Med. I.,
  RWTH Aachen,
  Goethestr. 27 / FRG.

Dr. K. Mohnike,
  1040 Berlin,
  Hessische Str. 3/4,
  Inst. Physiol.,
  Humboldt-Univ.,
  Berlin / GDR.

Dr. E. Musso,
  Parma,
  Via Gramsci 14,
  Ist. Physiol. Gen.
  Univ. Parma / Italy.

Dr. T. Palko,
  Sinai Hospital of Detroit,
  6767 West Outer Drive,
  Detroit MI 48235 / USA.

Dr. J. Préda,
  H. 1389 Budapest XIII,
  Szabolcs u. 35,
  2nd Medical Clinic,
  Postgraduate Medical School /
  Hungary.

Dr. J. Przybylski,
  Sinai Hospital of Detroit,
  6767 West Outer Drive,
  Detroit MI 48235 / USA.

Dr. G. Rau,
  Aachen,
  Goethestr. 27,
  Helmholtz-Inst. Biomed. Eng.,
  RWTH Aachen /FRG.

Dr. G. Regoliosi,
  Roma,
  Viale del Policlinico 137,
  Ist. Applic. del Calcolo,
  CNR / Italy.

Prof. Dr. P. Rijlant,
  1000 Brussels,
  115 Boulevard de Waterloo,
  Inst. Solvay / Belgium.

Dr. A. Rolli,
  Parma,
  Ospedale Maggiore,
  Divisiona di Cardiologia /
  Italy.

Prof. Dr. Roshchevsky,
  167610 Syktyvkar,
  Kommunistitcheskaja 24,
  Inst. Biol.,
  Acad. Sci. USSR / USSR.

Dr. M. Rubenfire,
  Sinai Hospital of Detroit,
  6767 West Outer Drive,
  Detroit MI 48235 / USA

Prof. Dr. I. Ruttkay-Nedecký,
  884 23 Bratislava,
  Sienkieviczova 1,
  Inst. Physiol. SAV / CSSR.

Dr. A. Ruttkay-Nedecká,
  884 23 Bratislava,
  Sienkieviczova 1,
  Inst. Physiol. SAV / CSSR.

Prof. Dr. G. Schoffa,
  7500 Karlsruhe,
  Lehrgebiet Biophysik,
  Univ. Karlsruhe / FRG.

Prof. Dr. E. Schubert,
  1040 Berlin,
  Hessische Str. 3/4,
  Physiol. Inst.,
  Humboldt-Univ.
  Berlin / GDR.

Dr. H. Schwartze,
  7010 Leipzig,
  Liebigstr. 27,
  Chair of Pathophysiology,
  Bereich Medizin,
  Karl-Marx-University,
  Leipzig / GDR.

Dr. G. Sedek,
  Sinai Hospital of Detroit,
  6767 West Outer Drive,
  Detroit MI 48235 / USA.

Dr. J. Silny,
  Aachen,
  Goethestr. 27,
  Hemholtz-Inst. Biomed. Eng.,
  RWTH Aachen / FRG.

Dr. S. Spaggiari,
  Parma,
  Via Gramsci 14,
  Ist. Physiol. Gen.
  Univ. di Parma / Italy.

Dr. D. Stilli,
  Parma,
  Via Gramsci 14,
  Ist. Physiol. Gen.,
  Univ. di Parma / Italy.

Dr. V. Szathmáry,
  884 23 Bratislava,
  Sienkieviczova 1,
  Inst. Physiol. SAV / CSSR.

Prof. Dr. B. Taccardi,
  Parma,
  Via Gramsci 14,
  Ist. Gen. Physiol.,
  Univ. Parma / Italy.

Dr. P. Tekel,
  885 27 Bratislava,
  Dúbravská cesta,
  Ustav merania SAV / CSSR.

Dr. L. I. Titomir,
  101447 Moscow,
  Ermolovoy 19,
  Inst. of the Problems of
  Information Transmission,
  Acad. Sci. USSR / USSR.

Dr. C. Viganotti,
  27100 Pavia,
  Corso Carlo Alberto 5,
  Lab. di analisi num.
  CNR / Italy.

Dr. W. Wajsczcuk,
  Sinai Hospital of Detroit,
  6767 West Outer Drive,
  Detroit MI 48235 / USA.

Dr. A. Whitty,
  Sinai Hospital of Detroit,
  6767 West Outer Drive,
  Detroit MI 48235 / USA

Dr. R. Zochowsky,
  Sinai Hospital of Detroit,
  6767 West Outer Drive,
  Detroit MI 48235 / USA.

INDEX

227